Springer Complexity

Springer Complexity is an interdisciplinary program publishing the best research and academic-level teaching on both fundamental and applied aspects of complex systems – cutting across all traditional disciplines of the natural and life sciences, engineering, economics, medicine, neuroscience, social and computer science.

Complex Systems are systems that comprise many interacting parts with the ability to generate a new quality of macroscopic collective behavior the manifestations of which are the spontaneous formation of distinctive temporal, spatial or functional structures. Models of such systems can be successfully mapped onto quite diverse "real-life" situations like the climate, the coherent emission of light from lasers, chemical reaction-diffusion systems, biological cellular networks, the dynamics of stock markets and of the internet, earthquake statistics and prediction, freeway traffic, the human brain, or the formation of opinions in social systems, to name just some of the popular applications.

Although their scope and methodologies overlap somewhat, one can distinguish the following main concepts and tools: self-organization, nonlinear dynamics, synergetics, turbulence, dynamical systems, catastrophes, instabilities, stochastic processes, chaos, graphs and networks, cellular automata, adaptive systems, genetic algorithms and computational intelligence.

The three major book publication platforms of the Springer Complexity program are the monograph series "Understanding Complex Systems" focusing on the various applications of complexity, the "Springer Series in Synergetics", which is devoted to the quantitative theoretical and methodological foundations, and the "SpringerBriefs in Complexity" which are concise and topical working reports, case-studies, surveys, essays and lecture notes of relevance to the field. In addition to the books in these two core series, the program also incorporates individual titles ranging from textbooks to major reference works.

Understanding Complex Systems

Founding Editor: S. Kelso

Future scientific and technological developments in many fields will necessarily depend upon coming to grips with complex systems. Such systems are complex in both their composition – typically many different kinds of components interacting simultaneously and nonlinearly with each other and their environments on multiple levels – and in the rich diversity of behavior of which they are capable.

The Springer Series in Understanding Complex Systems series (UCS) promotes new strategies and paradigms for understanding and realizing applications of complex systems research in a wide variety of fields and endeavors. UCS is explicitly transdisciplinary. It has three main goals: First, to elaborate the concepts, methods and tools of complex systems at all levels of description and in all scientific fields, especially newly emerging areas within the life, social, behavioral, economic, neuro- and cognitive sciences (and derivatives thereof); second, to encourage novel applications of these ideas in various fields of engineering and computation such as robotics, nano-technology and informatics; third, to provide a single forum within which commonalities and differences in the workings of complex systems may be discerned, hence leading to deeper insight and understanding.

UCS will publish monographs, lecture notes and selected edited contributions aimed at communicating new findings to a large multidisciplinary audience.

More information about this series at http://www.springer.com/series/5394

Sven Banisch

Markov Chain Aggregation for Agent-Based Models

 Springer

Sven Banisch
Max Planck Institute for Mathematics
    in the Sciences
Leipzig
Germany

ISSN 1860-0832   ISSN 1860-0840   (electronic)
Understanding Complex Systems
ISBN 978-3-319-79691-8   ISBN 978-3-319-24877-6   (eBook)
DOI 10.1007/978-3-319-24877-6

Springer Cham Heidelberg New York Dordrecht London
© Springer International Publishing Switzerland 2016
Softcover reprint of the hardcover 1st edition 2016

Printed on acid-free paper

Springer International Publishing AG Switzerland is part of Springer Science+Business Media
(www.springer.com)

*Die Mathematik ist eine
Karikatur der Wirklichkeit*

(Philippe Blanchard)

Preface

Agent-based modeling is an interesting tool. It provides model developers with a great degree of freedom for the design of systems in which heterogeneous entities interact with each other and the environment. Agent-based models (ABMs) are therefore a great tool to explore how different assumptions about how individuals behave and interact affect the evolution of social, economic or ecological systems as a whole.

The mathematical formalization of these models, however, is still in its infancy partly due to the fact micro-level heterogeneity or complex interaction structures often lead to effects in the system dynamics which are not easily accounted for by macroscopic formulations of the problem. To address this issue an understanding of the transition from the most informative "atomic" level to the levels at which the system behavior is typically observed is important, because it can help to derive and evaluate models on specific levels, on the one hand, and to understand the temporal and spatial patterns that may emerge in that transition on the other. The book at hand develops a Markov chain approach that allows a rigorous analysis of a class of microscopic models which specify the dynamics of a complex system at the individual level. It provides a general framework of aggregation in agent-based and related computational models by making use of lumpability and information theory in order to link between the micro and macro levels of observation.

The starting point is a microscopic Markov chain description of the dynamical process in complete correspondence with the dynamical behavior of the ABM, which is obtained by considering the set of all possible agent configurations as the state space of a huge Markov chain. This is referred to as micro chain, and an explicit formal representation including microscopic transition rates can be derived for a class of models by using the random mapping representation of a Markov process. The explicit micro formulation enables the application of the theory of Markov chain aggregation—namely, lumpability—in order to reduce the state space of the micro chain and relate microscopic descriptions to a macroscopic formulation of interest. Well-known conditions for lumpability make it possible to establish the cases where the macro model is still Markov, and in this case we obtain a complete picture of the dynamics including the transient stage, the most interesting phase in applications.

For such a purpose a crucial role is played by the type of probability distribution used to implement the stochastic part of the model which defines the updating rule and governs the dynamics. Namely, if we decide to remain at a Markovian level, then the partition, or equivalently, the collective variables, used to build the macro model must be compatible with the symmetries of the probability distribution ω. Microscopic heterogeneity and constraints translate into dynamical irregularities in the micro chain and require a refinement of the aggregation and the corresponding level of observation. This underlines the theoretical importance of homogeneous or complete mixing in the analysis of "voter-like" models at use in population genetics, evolutionary game theory and social dynamics.

The problem of aggregation in ABMs and the lumpability conditions in particular can be embedded into a more general framework which makes use of information theory in order to identify different levels and relevant scales in complex dynamical systems. Lumpability and, respectively, the existence of a higher-level Markovian description is one out of several mutually related criteria which a closed higher-level description should satisfy. Consequently, the application of information-theoretic measures of closure to ABMs allows us to quantify the information that is lost in the transition from the micro dynamics to a particular macro description. The method informs us in this way about the complexity of a system introduced by nontrivial interaction relations. Namely, if a favored level of observation is not compatible with the symmetries in ω, a certain amount of memory is introduced by the transition from the micro level to such a macro description, and this is the fingerprint of emergence in ABMs. The resulting divergence from Markovianity can be quantified using information theory, and the book presents a scenario in which different closure measures can be explicitly computed.

Throughout the book, we mainly rely on two simple models to illustrate these theoretical ideas: the voter model (VM) and an extension of it called the contrarian voter model (CVM). Using these examples, the book shows that Markov chain theory allows for a rather precise understanding of the model dynamics in case of "simple" population structures where a tractable macro chain can be derived. Constraining the system by interaction networks with a strong local structure leads to the emergence of meta-stable states in the transient of the model. Constraints on the interaction behavior such as bounded confidence or assortative mating lead to the emergence of new absorbing states in the associated macro chain and are related to stable patterns of polarization (stable coexistence of different opinions or species). Constraints and heterogeneities in the microscopic system and complex social interactions are the basic characteristics of ABMs, and the Markov chain approach to link the micro chain to a macro-level description (and likewise the failure of a Markovian link) highlights the crucial role played by those ingredients in the generation of complex macroscopic outcomes.

This book has developed out of my dissertation project at the department of Mathematical Physics at the University of Bielefeld. I am very grateful to my supervisor Philippe Blanchard and to Dima Volchenkov (both in Bielefeld) for an open ear whenever I knocked on their doors and for the free environment they provided. I am also very grateful to Ricardo Lima (Marseilles), who is probably the person who engaged most in the details of this project, and to Tanya Araújo (Lisbon) for her advice and encouragement. Chapter 8 has been developed in cooperation with Tanya. Especially the parts dealing with the application of information-theoretic measures have benefited a lot from discussions with Eckehard Olbrich and Robin Lamarche-Perrin (both in Leipzig). All of this would have been a lot more difficult without the unconditional support of my family.

Finally, I gratefully acknowledge financial support of the German Federal Ministry of Education and Research (BMBF) through the project *Linguistic Networks* (http://project.linguistic-networks.net) and the European Community's Seventh Framework Programme (FP7/2007-2013) under grant agreement no. 318723 (*MatheMACS* http://www.mathemacs.eu). Both projects have provided a very inspiring environment. Financial support by the Klaus Tschira Foundation (http://www.klaus-tschira-stiftung.de) during the finalization of this book is also gratefully acknowledged.

Leipzig, Germany   Sven Banisch

Contents

Chapter 1
Introduction

1.1 Complex Multi-Level Systems

I think that nowadays most people would confirm that the world we live in is a complex one. Not only the problems that we face at a global scale (such as climate change and financial crises) but also many of our very personal day-to-day decisions (such as choosing between a fresh organic apple from oversee and a local apple maintained in an energy-expensive cooling chamber) involve nowadays, if carefully considered, the evaluation of entanglements of global scope. There is a high level of uncertainty in the evaluation of the consequences of our actions owing to the fact that those entanglements are often not clearly evident. There is also a high degree of freedom in what concerns the number of options that are in principle at our disposal, but if we do not sufficiently understand the functioning of the system there is no way to choose among them.

The "new science of complex systems" is an attempt to better understand the behavior of systems that are composed of elementary units and structures of mutual dependencies (*Wechselwirkungen*) between those units. The fundamental idea is that *complex patterns of higher-level organization emerge in a dynamical system of interacting individuals that participate in a self-organizing process.* While no central control is assumed to direct this process, the global emergences that are generated by it may well have an effect on the individual dynamics. Complexity, in this dynamical context, relates to the fact that higher-level patterns and processes are not easily understood by considering the dynamical mechanisms at the lower level only.

Of course, the fact that the behavior of many real-world systems is not predictable in simple way from the behavior of the system's components has been acknowledged long ago. Likewise, the observation that systems from very different fields and at different scales share important principles of organization. But especially the last two decades have witnessed a tremendous increase in scientific activity trying to make visible the empirical fingerprints of complex behavior (such as power

© Springer International Publishing Switzerland 2016
S. Banisch, *Markov Chain Aggregation for Agent-Based Models*,
Understanding Complex Systems, DOI 10.1007/978-3-319-24877-6_1

law distributions or long range correlations) on the one hand, and to extract the underlying mechanisms and causal relations at work in those systems in order to really understand the fundamental principles of self-organized complexity on the other. For its enormous range of application—from biology to sociology, from physics to linguistics—complexity has become one of the most promising concepts in modern science.

In all of this, computational tools have become very important. Several methodological innovations are in fact enabled only by the general availability of relatively powerful computers: from the retrieval of information, statistical regularities and patterns from large amounts of data to the simulative testing of different behavioral assumptions in models of biological, social or cultural evolution. In general, the use of computational instruments does not make mathematics dispensable, to the contrary, it rather calls for the development of sound mathematical foundations of these new methods. In data science, this relates to questions concerned with statistical significance, algorithmic complexity, information theory, among many others; for computational models, it is related to proper formal model specifications, to the development of mathematical theories for multi-level systems and analytical solution strategies.

This volume is concerned with the latter problem area. It develops mathematical concepts for the formal treatment of a class of computational models. Namely, it formulates agent-based models—models in which a finite number of agents interact according to simple behavioral assumptions—as Markov chains and makes use of Markov chain theory to derive explicit statements about the possibility of linking a microscopic agent model to the dynamical processes at the level of macroscopic observables. The questions that are addressed in this book are inherently dynamic ones: the focus is not on the structural properties of certain agent networks, but rather on the dynamical processes at the micro and the macro level that differently structured systems give rise to. A particular aspect in that is the role that microscopic heterogeneity and constraints in the agent behavior play in the generation of macroscopic complexity. In this way, the book touches upon questions related to the micro-macro link in social simulation and to computational emergence in general. Moreover, the question of deriving macroscopic descriptions with a minimal loss of information also goes to the heart of statistical mechanics.

1.2   Microsimulation and Agent-Based Models

Recent improvements in multidisciplinary methods and, particularly, the availability of powerful computational tools are giving researchers an ever greater opportunity to investigate societies in their complex nature. The adoption of a complex systems approach allows the modeling of macro-sociological or economic structures from a bottom-up perspective—understood as resulting from the repeated local interaction of socio-economic agents—without disregarding the consequences of the structures themselves on individual behavior, emergence of interaction patterns and social welfare.

Agent-based models (henceforth ABMs) are at the leading edge of this endeavor. ABMs are an attempt to understand how macroscopic regularities may emerge through processes of self-organization in systems of interacting agents. The main idea is to place a population of agents characterized by a set of attributes within a virtual environment and specify simple rules of how agents interact with each other and the environment. The interaction rules are usually based on simple behavioral assumptions with the aim to mimic the individual behavior of real actors in their local environment. While the system is modeled at the microscopic level, its explanatory scope is the macro level. In that, AB modeling follows the tradition of methodological individualism which claims "that social phenomena must be explained by showing how they result from individual actions" (Heath 2011, par.1).

AB systems are dynamical systems. Typically implemented on a computer, the time evolution is computed as an iterative process—an algorithm—in which agents are updated according to the specified rules. ABMs usually also involve a certain amount of stochasticity, because the agent choice and sometimes also the choice among different behavioral options is random. This is why Markov chain theory is such a good candidate for the mathematical formalization of ABMs.

The Voter Model (VM from now on) is a simple paradigmatic example (Kimura and Weiss 1964; Castellano et al. 2009, among many others). In the VM, agents can adopt two different states, which we may denote as white $\square$ and black $\blacksquare$. The attribute could account for the opinion of an agent regarding a certain issue, its approval or disapproval regarding certain attitudes. In an economic context, $\blacksquare$ and $\square$ could encode two different behavioral strategies, or, in a biological context, the occurrence of mutants in a population of individuals. The iteration process implemented by the VM is very simple. At each time step, an agent i is chosen at random along with one of its neighboring agents j and one of them imitates the state of the other. In the long run, the model leads to a configuration in which all agents have adopted the same state (either $\square$ or $\blacksquare$). In the context of biological evolution, this has been related to the fixation or extinction of a mutant in a population. The VM has also been interpreted as a simplistic form of a social influence process by which a shared convention is established in the entire population.

Let us consider an example simulation run of the VM to provide an intuition about its behavior (Fig. 1.1). Assume there are 20 agents connected by a chain such that an agent at position i is connected to agents $i - 1$ and $i + 1$ (except the first and the last agent who have only one neighbor). Let the random initial population be $\mathbf{x} = ($ ■■■□■■■■■□□□■□■□■■□□■ $)$ corresponding to the left-most column in Fig. 1.1. The time evolution is shown from left to right, the columns represent the configuration of the population each time after ten VM steps have been performed. This example shows two main features of the VM: (1) the emergence of a meta-stable transient state of local alignment, and (2) the final convergence to complete consensus. The first feature is clearly due to the interaction topology because initial local divergences are leveled with a high probability and once an areal of local alignment is achieved change is admitted, due to the chain topology,

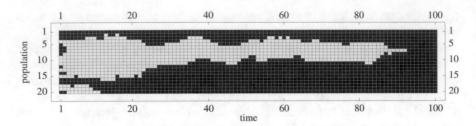

Fig. 1.1 Example of the time evolution of the VM on the chain network

only at the borders of that domain. The second feature is a more general feature of the finite VM: sooner or later consensus occurs in every connected topology.

When designing an agent model, one is inevitably faced with the problem of finding an acceptable compromise between realism and simplicity. If many aspects are included into the agent description, the model might be plausible with regard to the individual behaviors, but it will be impossible to derive rigorous analytical results. In fact, it can even be very hard to perform systematic computations to understand the model dynamics if many parameters and rules are included. On the other hand, models that allow for an analytical treatment often oversimplify the problem at hand. The VM is good example of this kind. In AB modeling, we can find the whole spectrum between these two extremes. While simplicity is often favored by physicists in order to be able to apply their well-developed tools from statistical physics, more realistic descriptions are often desired by researchers in the humanities because they are interested in incorporating into the model a reasonable part of their qualitative knowledge at the micro and macro scales. Both views have, of course, their own merits.

1.3 Markov Chain Description of Agent-Based Models

This work is a contribution to interweaving two lines of research that have developed in almost separate ways: ABMs and Markov chains. The latter represents the simplest form of a stochastic process while the former puts a strong emphasis on heterogeneity and social interactions. The main expected output of a Markov chain strategy applied to AB systems is a better understanding of the relationship between microscopic and macroscopic dynamical properties. Moreover, we aim to contribute not only to the understanding of the asymptotic properties of ABMs but also to the transient mechanisms that rule the system on intermediate time scales. For practical purposes this is the most relevant information for two reasons: first, in many cases the chains are absorbing, so the asymptotic dynamics is trivial and second, they describe the evolution of the system before external perturbations take place and possibly throw it into a new setting.

The possibility of using Markov chains in the analysis of ABMs has been pointed out in Izquierdo et al. (2009). The main idea is to consider all possible configurations of the agent system as the state space of a huge Markov chain. While Izquierdo et al. (2009) mainly rely on numerical computations to estimate the stochastic transition matrices of the models, here we show for a class of models how to derive explicitly the transition probabilities $\hat{P}$ in terms of the update function $\mathbf{u}$ and a probability distribution ω accounting for the stochastic parts in the model. It turns out that realizations of ABMs with a sequential update scheme can be conceived as random walks on regular graphs.

Consider an AB system defined by a set $\mathbf{N}$ of agents, each one characterized by individual attributes that are taken from a finite list of possibilities. We denote the set of possible attributes by $\mathbf{S}$ and we call the *configuration space* Σ the set of all possible combinations of attributes of the agents, i.e. $\Sigma = \mathbf{S}^N$. Therefore, we denote an *agent configuration* as $\mathbf{x} \in \Sigma$ and write $\mathbf{x} = (x_1, \ldots, x_i, \ldots, x_N)$ with $x_i \in \mathbf{S}$. The updating process of the attributes of the agents at each time step typically consists of two parts. First, a random choice of a subset of agents is made according to some probability distribution ω. Then the attributes of the agents are updated according to a rule $\mathbf{u}$, which depends on the subset of agents selected at this time. With this specification, ABMs can be represented by a so-called random map representation which may be taken as an equivalent definition of a Markov chain (Levin et al. 2009). We refer to the process $(\Sigma, \hat{P})$ as *micro chain*.

1.4 Markov Chain Aggregation

When performing simulations of an ABM we are actually not interested in all the dynamical details but rather in the behavior of certain macro-level properties that inform us about the global state of the system (such as average opinion, number of communities, etc.). The explicit formulation of ABMs as Markov chains enables the development of a mathematical framework to link a micro chain corresponding to an ABM to such a macro-level description of interest. Namely, from the Markov chain perspective, the transition from the micro to the macro level is a projection of the micro chain with state space Σ onto a new state space $\mathbf{X}$ by means of a (projection) map Π from Σ to $\mathbf{X}$. The meaning of the projection Π is to lump sets of micro configurations in Σ into an aggregate set according to the macro property of interest. Such a situation naturally arises if the ABM is observed not at the micro level of Σ, but rather in terms of a measure ϕ on Σ by which all configuration in Σ that give rise to the same measurement are mapped into the same macro state, say $X_k \in \mathbf{X}$. An illustration of such a projection is provided in Fig. 1.2.

There are two things that may happen when projecting a micro process onto a macroscopic state space $\mathbf{X}$. First, under certain conditions the macro-level process is still a Markov chain. This case is known as *lumpability* in Markov chain theory and necessary and sufficient conditions are provided in a well-known textbook on finite Markov chains by Kemeny and Snell (1976). The questions addressed in what

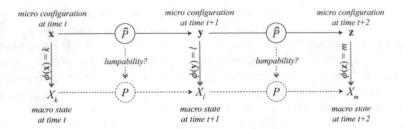

Fig. 1.2 A micro process ($\mathbf{x}, \mathbf{y}, \mathbf{z} \in \Sigma$) is observed ($\phi$) at a higher level and this observation defines another macro level process ($X_k, X_l, X_m \in \mathbf{X}$). The micro process is a Markov chain with transition matrix $\hat{P}$. The macro process is a Markov chain (with P) only in the case of lumpability

follows concerns, first of all, the conditions on the microscopic system and the projection construction that have to be met in order to lead to a macro process that is still a Markov chain. In this regard, if we decide to remain at a Markovian level, then the projection, or equivalently the collective variables to be used to build the macro model must be compatible with the symmetry of the probability distribution ω. In turn, in the absence of any symmetry, there is no other choice than to stay at the micro-level because no Markovian macro-level description is possible in this case.

Secondly, and more generally, the price to pay in passing from the micro to the macro dynamics by such a projection construction is that the projected system is no longer a Markov chain. Long memory (even infinite) may appear in the projected system. Consequently, this setting can provide a suitable framework to understand how aggregation may lead to the emergence of long range memory effects. This opens up a series of interesting questions: for instance, why and in what sense does the behavior of the macro process deviate from Markovianity? How can we measure these deviations? Do we introduce memory or long-range correlations at the macro level by the very way we observe a process and is the emergence of these effects just due to an aggregation which is insensitive to microscopic heterogeneities? In particular, there is usually a strong interest in the effects that different interaction topologies have on the transient model dynamics as well as on the emergence of characteristic meta-stable situations, such as the persistent pattern of local alignment shown in Fig. 1.1. In that regard, how good does the mean field solution approximate network dynamics and for which networks does it provide acceptable approximations? Is there an alternative macro-level formulation that leads to better results? If yes, which properties can be captured by it? A micro-macro formalism may shed new light on some of these questions.

To my point of view, the non-Markovian case is in many ways even more interesting than the case of lumpability. In particular, because it relates microscopic heterogeneity to macroscopic complexity (structure generation). Constraints, heterogeneities in the microscopic system and complex social interactions are the basic characteristics of ABMs, and the Markov chain approach to link the micro chain to a macro level description (and likewise the failure of a Markovian link) highlights the crucial role played by those ingredients in the generation of complex macroscopic outcomes. The formalization of the relations between the micro and the macro levels in the description of the dynamics of AB systems as well as their mathematical characterization is a step towards a mathematical theory of emergence in complex adaptive systems.

To address these issues, the book includes a chapter which applies recently developed information-theoretic measures for multi-level systems to quantify deviations from Markovianity introduced through the transition from one level of description to the other (see Chap. 7). It shows that memory effects are introduced by a global aggregation over the agent population without sensitivity to micro- or mesoscopic structures. While lumpability is a yes-no question, the information-theoretic setting provides understanding about the amount of information that is lost in the transition from the micro to the macro level.

1.5 Micro-Macro Transition in the Voter Model

Let us exemplify the link between a micro and a macro chain by Markov chain aggregation for the VM. From the microscopic perspective, the VM corresponds to an absorbing random walk on the N-dimensional hypercube. If N agents can be in two different states, the set of all agent configurations Σ is the set of all bit-strings of length N. Due to the dyadic conception of the interaction along with a sequential update scheme only one agent may change at a time which means that transitions are only possible between configurations that differ in at most one bit. The structure of the VM micro chain is shown for a small system of three agents in the upper part of Fig. 1.3.

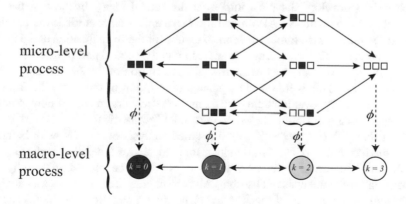

Fig. 1.3 Micro and macro level in the VM with three agents

In the VM, the most typical level of observation is to count the number of agents in the different states. In hypercube terminology this corresponds to the Hamming weight (i.e., $\phi(\mathbf{x}) = h(\mathbf{x})$). By the projection that this observation induces, all micro configurations with the same number of (say) white agents are mapped into the same macro state. If k is the number of white agents ($h(\mathbf{x}) = k$), we denote the respective macro state as X_k. Therefore, if we are dealing with a system of N agents, there are $N + 1$ macro states which is a tremendous reduction compared to the 2^N micro configurations. The projection construction for the VM is shown in Fig. 1.3.

Voter-like models—as used in physics-inspired models of social dynamics as well as in population genetics or evolutionary dynamics—are nice examples where such a projection construction is particularly meaningful. Namely, because it corresponds to the most typical description of the model dynamics in terms of attribute frequencies. Lumpability allows to determine conditions for which the macro chain on $\mathbf{X} = (X_0, \ldots, X_k, \ldots, X_N)$ is again a Markov chain and, as will be shown in Chap. 3, this requires that the probability distribution ω over agent choices must be invariant under the group $\mathscr{S}_N$ of all the permutations of N agents, and therefore uniform. This underlines the theoretical importance of homogeneous mixing and respectively the complete graph in the analysis of the VM and related models.

1.6 Outline

The book is organized into ten chapters. Chapter 2 provides an overview over AB modeling, their mathematical formalization as Markov chains as well as other concepts that will play an important role in the remainder. It also reviews different approaches to lumpability in Markov chains and motivates their application to ABMs.

Chapter 3 develops the most important theoretical ideas. In the first part, an elementary introduction to ABMs and their dynamical characteristics is provided which illustrates the usefulness of Markov chains for the formalization of the micro-level dynamics. This formalization is addressed in the second part of Chap. 3, which shows how to derive an exact Markov chain description of the AB dynamics for a class of models. This is followed by general description of the transition from the micro to the macro level including the main theoretical arguments of how to derive lumpable macro-descriptions based on the model symmetries.

After that, Chap. 4 proceeds with a detailed analysis of the VM with homogeneous mixing and applies Markov chain tools to obtain a detailed understanding of the transient model dynamics. This analysis may be read as a step-by-step instruction for the analysis of absorbing Markov chains. It also includes an analysis of the multi-state version of the VM and shows that interaction constraints such as bounded confidence may lead to the stable co-existence of polarization.

In Chap. 5, we discuss what happens in the case of inhomogeneous interaction probabilities. A systematic approach to aggregation is developed which exploits all the dynamical redundancies that have its source in the agent network on which the model is implemented. This enables the analytical treatment of a leader-follower system as well as the two-community model in which two strongly connected groups interact weakly with one another.

Chapter 6 applies the results of the previous chapters to an extension of the VM that accounts for non-conformity behavior. As the model introduces a small probability with which agents act as contrarians opposing the state of their interaction partner, this model is called contrarian voter model (CVM). Starting from the corresponding micro-level description, we analyze the behavior of this model on the complete and the two-community graph using Markov chain tools. Since the CVM leads to a non-absorbing, regular Markov chain, this analysis provides an overview over some of the instruments available for the exploration of non-absorbing chains.

The main objective of Chap. 7 is to study the influence of interaction topology on the macro-level dynamics in the case of non-lumpability and to apply information-theoretic tools to better understand such situations. After a discussion of these tools and how they relate to lumpability, the two-community CVM is used as an analytical scenario to study the discrepancy between the lumpable homogeneous mixing case and the model on a slightly more complex topology. The possibility of weak lumpability is also discussed.

Chapter 8 shows how the framework developed throughout this book can be used to understand the (sometimes very subtle) differences between different implementations of AB systems that are designed on the basis of very similar conceptual ideas. We inspect two well-known models with application in population genetics on the one and social dynamics on the other hand. While one main focus in the former is on the adaptive behavior of a population, the latter is more often concerned with the emergence of polarization and stable clusters of individuals. Even though the dynamical update rules used in the modeling of the microscopic interactions follow the same principles, the behavior of the two models differs fundamentally which we illustrate by numerical simulations. A Markov chain analysis of the respective minimal model variants then reveals that the reason for these differences is due to the way the mechanisms of agent selection, interaction and replacement are constrained and combined in the modeling.

While the specific issues are discussed at the end of each chapter, Chap. 9 aims at a synthetic view on how this work may contribute more generally to the study of complexity and emergence. A provisional definition of emergence in terms of lumpability provides a link between two different perspectives on emergence, namely, the concept of dynamical incompressibility and Wimsatt's notion of non-aggregativity. Finally, Chap. 10 draws a conclusion on the project as a whole and outlines some ideas and challenges for future research.

References

Castellano, C., Fortunato, S., & Loreto, V. (2009). Statistical physics of social dynamics. *Reviews of Modern Physics, 81*(2), 591–646.

Heath, J. (2011). Methodological individualism. In E. N. Zalta (Ed.), *The Stanford encyclopedia of philosophy* (Spring 2011 edn.). http://plato.stanford.edu/cgi-bin/encyclopedia/archinfo.cgi?entry=methodological-individualism.

Izquierdo, L. R., Izquierdo, S. S., Galán, J. M., & Santos, J. I. (2009). Techniques to understand computer simulations: Markov chain analysis. *Journal of Artificial Societies and Social Simulation, 12*(1), 6.

Kemeny, J. G., & Snell, J. L. (1976). *Finite Markov chains*. New York: Springer.

Kimura, M., & Weiss, G. H. (1964). The stepping stone model of population structure and the decrease of genetic correlation with distance. *Genetics, 49*, 561–576.

Levin, D. A., Peres, Y., & Wilmer, E. L. (2009). *Markov chains and mixing times*. Providence: American Mathematical Society.

Chapter 2
Background and Concepts

This work is a contribution to interweaving two lines of research that have developed in almost separate ways: Markov chains and agent-based models (ABMs). The former represents one of the simplest forms of a stochastic process while the latter puts a strong emphasis on heterogeneity and social interactions. This chapter provides an introduction to AB modeling and reviews approaches to use Markov chains in their analysis.

The main expected output of the Markov chain strategy applied to ABMs is a better understanding of the relationship between microscopic and macroscopic dynamical properties. This brings into the discussion concepts of aggregation and emergence, and it also relates to macroscopic mean-field formulations as a substantial tool in the statistical mechanics approach to social dynamics. Moreover, a series of information-theoretic tools to put the notion of levels onto mathematical grounds have been developed in recent years. A complete review of the literature dealing with these topics is clearly beyond the scope of this chapter which is rather aimed at introducing the most important concepts with reference to AB systems and Markov chains. Especially the physics-inspired approach to social dynamics has attracted a lot of interest in the last years and a huge number of papers is still produced every year. For a relatively coherent review (though, may be, no longer completely up-to-date), the reader may be referred to Castellano et al. (2009).

2.1 Agent-Based and Related Models

ABMs are an attempt to understand how macroscopic regularities may emerge through processes of self-organization in systems of interacting agents. A system at question is modeled at the microscopic level by specifying the elementary units of that system—the agents—and implementing simple rules for how these agents interact with one another. Typically implemented on a computer, the time evolution

© Springer International Publishing Switzerland 2016
S. Banisch, *Markov Chain Aggregation for Agent-Based Models*,
Understanding Complex Systems, DOI 10.1007/978-3-319-24877-6_2

of such a system is computed as an iterative process—an algorithm—in which agents are updated according to the specified rules. One of the main purposes of this modeling strategy is "to enrich our understanding of fundamental processes" (Axelrod 1997, p. 25) underlying certain observed patterns, or to "explore the simplest set of behavioral assumptions required to generate a macro pattern of explanatory interest" (Macy and Willer 2002, p. 146).

One paradigmatic example of ABMs is Reynolds model of the flocking behavior of birds (Reynolds 1987). While the modeling of a flock as such is difficult, quite realistic flocking behavior is achieved if the individual birds follow simple rules of how to react upon the action of other individuals in their neighborhood. Another well-known example is Schelling's model of segregation (Schelling 1971). Here, two kinds of householders (say black and white) located on a lattice are endowed with a slight preference to settle in a neighborhood with more households of the same kind. Running that system leads to a clear spatial segregation at the global level even if the homophily preference is small. Similar effects can be observed in models of opinion and cultural dynamics, see, for instance, Axelrod (1997), Deffuant et al. (2001), Hegselmann and Krause (2002), and Banisch et al. (2010). Another paradigmatic problem that has been addressed by AB research is the emergence of a set of norms or common conventions. In the naming game proposed by Steels (1997), for instance, robots learn common word-object relations in a communication process based on trail and error. Other models in which an initial plurality in a population of agents evolves to a common consensus state include various models of opinion formation with the VM as the most simple representative (see Castellano et al. 2009 for a review of these models).

It is common to trace back the history of AB simulation to the cellular automata (henceforth CA) designed by von Neumann (1951) and later shaped by Berlekamp et al. (1982) and Wolfram (1983, 2002). And in fact, many ABMs can be viewed as a stochastic CA with asynchronous update. The methods developed in this work apply precisely to that type of models.

However, even some years before von Neumann and Ulam came up with the first CA design, another type of "individual-based" model had been introduced in a branch of theoretical biology which is today called population genetics (see Li 1977 for a collection of the seminal papers in that field). Wright and Fisher (along with Haldane known as the founders of population genetics) advocated a simple model for the evolution of allele frequencies (Wright 1932) based on microscopic assumptions of gene transmission from the parent to the children generation. In 1958, Moran (1958) made use of Markov chain theory to study a modified model and introduced what today is known as the Moran process. Later, Kimura went further in this line of research on a neutral theory of evolution with the stepping stone model (Kimura and Weiss 1964) which still later became known as the voter model (abbreviated by VM throughout this book). From the very beginning population genetics developed as a mathematical discipline and has inspired various solution strategies from probabilistic methods including Markov chains and coalescing random walks to mean field approaches in statistical physics.

The biological literature on evolutionary dynamics on graphs has mainly started from the model proposed by Moran (1958). In the Moran model, at each time step, an individual is chosen at random to reproduce and replaces a second one chosen at random as well. In the original model, there is no population structure which means that all individuals are chosen with equal probability. Therefore—this is something that will be made explicit in the fourth chapter of this thesis—the dynamics can be formulated as a birth-death random walk on the line. See Claussen and Traulsen (2005), Traulsen et al. (2005), and Nowak (2006) for treatments of the associated Moran process. While early studies (Maruyama 1974; Slatkin 1981) had indicated that population structure has no or only little effect on the model behavior, it has recently been shown that population structure can have a significant influence (Liberman et al. 2005; Nowak 2006; Shakarian et al. 2012; Voorhees and Murray 2013; Voorhees 2013, among many others). The setting—sometimes referred to as evolutionary graph theory (Liberman et al. 2005)—is usually as follows: suppose the is a population of N individuals with fitness 1; suppose that a mutant with fitness r is introduced in one of the individuals; what is the probability that the mutant invades the entire population? The Moran case of unstructured populations is usually taken as a benchmark such that a graph which leads to a fixation probability different from the unstructured case are said to suppress or respectively enhance selection.

In the physics literature, the analysis of binary models as the VM is usually based on mean-field arguments. The system dynamics is traced in form of an aggregate order parameter and the system is reformulated on the macro-scale as a differential equation which describes the temporal evolution of that parameter. In many cases, the average opinion (due to the analogy to spin systems often called "magnetization") has proven to be an adequate choice, but sometimes the number of (re)active interfaces yields a more handable transformation (e.g., Frachebourg and Krapivsky 1996; Krapivsky and Redner 2003; Vazquez and Eguíluz 2008). A mean-field analysis for the VM on the complete graph was presented by Slanina and Lavicka (2003), and naturally, we come across the same results using our method (Sect. 4.1.2). Slanina and Lavicka (2003) derive expressions for the asymptotic exit probabilities and the mean time needed to converge, but the partial differential equations that describe the full probability distribution for the time to reach the stationary state is too difficult to be solved analytically (Slanina and Lavicka 2003, p. 4). Further analytical results have been obtained for the VM on d-dimensional lattices (Cox 1989; Frachebourg and Krapivsky 1996; Liggett 1999; Krapivsky and Redner 2003) as well as for networks with uncorrelated degree distributions (Sood and Redner 2005; Vazquez and Eguíluz 2008). It is noteworthy, that the analysis of the VM (and more generally, of binary-state dynamics) on networks has inspired a series of solution techniques such as refined mean-field descriptions (e.g., Sood and Redner 2005; Moretti et al. 2012), pairwise approximation (e.g., De Oliveira et al. 1993; Vazquez and Eguíluz 2008; Schweitzer and Behera 2009; Pugliese and Castellano 2009) and approximate master equations (e.g., Gleeson 2011, 2013).

The early works in population genetics (Fisher 1930, in particular) have inspired still another modeling approach that is related to ABMs, namely, evolutionary game theory (see Smith 1982 for a seminal volume and Roca et al. 2009 for a recent review). Here, games are designed in which agents repeatedly play against one another adopting one out of a set of predefined strategies. A fitness is assigned to the combinations of strategies and the population evolves as a response to this fitness. As in the framework of statistical mechanics, the model evolution is typically captured in form of differential equation describing the evolution of the (relative) frequencies of the different strategies, referred to as replicator dynamics in this context (Taylor and Jonker 1978; Schuster and Sigmund 1983; Hofbauer and Sigmund 2003). One of the main purposes of this work is to spell out explicitly how to link the dynamics at the micro level to these macroscopic descriptions.

Finally, it is worth mentioning that research in economics has experienced a growing interest in modeling economic phenomena as the result of the interactions of heterogeneous individuals (Tesfatsion and Judd 2006). In particular in the field of finance, this has led to the development of ABMs for the identification of (macro) patterns of collective dynamics from (micro) investor heterogeneity in many financial settings (Cont and Bouchaud 2000; LeBaron 2000; Bornholdt 2001; Kaizoji et al. 2002; Hommes 2006; Preis et al. 2013; Krause and Bornholdt 2013; Patzelt and Pawelzik 2013). Noteworthy, there is also a number of empirical applications of Markov chains in the field of finance (e.g., Corcuera et al. 2005; Nielsen 2005; Norberg 2006). Interaction and heterogeneity on the one hand, and non-Gaussianity, heavy tails and long-range correlations on the other appear to be natural features of modern economies, to which the formerly dominating tradition of modeling representative agents has, to a large extent, paid little attention. This thesis shows that memory effects at the macroscopic level are an immediate consequence of microscopic heterogeneity and it may therefore contribute to the identification of the relevant microscopic mechanisms that presumably play a role in the market.

2.2 Markov Chain Formalization of Agent-Based Models

The AB approach is first and foremost a computational methodology and the mathematical formalization of the models is in its infancy. This is probably due to the fact that a major motivation in the development of AB simulation has been to relax a series of unrealistic assumptions made in other modeling frameworks just in order to keep mathematical tractability; namely, rationality, perfect information, agent homogeneity, and others. The other side of the coin is that the focus on computer models and algorithms makes difficult the comparison of different models and also complicates a rigorous analysis of the model behavior. In fact, the problems of code verification and model comparison including the discussion of standards for the replication of ABMs has nowadays become an area of research in its own (e.g., Axtell et al. 1996; Axelrod 2003; Hales et al. 2003; David et al. 2005; Grimm et al. 2006; Wilensky and Rand 2007; Galán et al. 2009). As a matter of fact, many of those problems would actually vanish with a sound mathematical formulation of an AB simulation model. On the other hand, it is also clear that the precise mathematical specification of a high-dimensional system of heterogeneous interacting agents along with their update mechanisms can be cumbersome in more complex models.

To the authors knowledge, the first systematic approach to the development of mathematical formalism for ABMs in general is due to Laubenbacher and co-workers. Laubenbacher et al. (2009) review existing formal frameworks that have the potential to model AB systems, such as cellular automata and finite dynamical systems and argue for the latter as an appropriate mathematical framework to represent ABMs. However, the probabilistic nature of most models can only be accounted for by the stochastic version—the so-called stochastic finite dynamical systems—the analysis of which "is still in its infancy" (Laubenbacher et al. 2009, p. 14). On the other hand, Laubenbacher et al. (2009) recognize that stochastic finite dynamical systems give rise to Markov chains. However, for reasons that do not become very clear in their paper, the authors argue:

> To understand the effect of structural components such as the topology of the dependency graph or the stochastic nature of the update, it is important to study them not as Markov chains but as SFDS [stochastic finite dynamical systems] (Laubenbacher et al. 2009, p. 10)

I clearly disagree with them in this point, because the microscopic specification of ABMs as Markov chains developed in this thesis turns out to be a useful starting point for further analysis. But of course, the incentive of Laubenbacher et al. (2009) to further elaborate the theory of stochastic dynamical systems in order to derive rigorous results for ABMs in future is highly appreciable.

The usefulness of the Markov chain formalism in the analysis of ABMs has first been realized by Izquierdo et al. (2009). The authors look at ten well-known social simulation models and discuss for each of them how to represent the model as a time-homogeneous Markov chain. Among the models studied in Izquierdo et al. (2009) are the Schelling segregation model (Schelling 1971, for which some analytical results are available, for example, in Pollicott and Weiss 2001; Grauwin et al. 2010), the Axelrod model of cultural dissemination (Axelrod 1997, see also Castellano et al. 2000 for a mean-field approximation) and the sugarscape model from Epstein and Axtell (1996). Noteworthy, the sugarscape model—one of the reference models in the field of social simulation—contains virtually all features that may occur in ABMs: heterogeneous agents placed in a dynamic spatial environment, death and birth of agents, various static and dynamic attributes that may evolve on different time scales.

The main idea of Izquierdo et al. (2009) is to consider all possible configurations of the system as the state space of a huge Markov chain and the construction of that state space is actually the main challenge for Izquierdo and co-workers. Despite the fact that all the information of the dynamics of the ABM is encoded in a Markov chain, however, it is difficult to learn directly from this fact, due to the huge dimension of the configuration space and its corresponding Markov transition matrix. The analyses provided in Izquierdo et al. (2009) are essentially based on the classification of states into transient and absorbing communicating classes which allows some statements about the convergence as times goes to infinity.

The paper of Izquierdo et al. (2009) is designated "for researchers who may not have a strong mathematical background" (par.1.1) and probably therefore lacks rigorous arguments sustaining some of the results. Most fundamentally, there is no proof that the process on the constructed configuration space indeed satisfies the Markov property. Their work also mainly relies on numerical computations to estimate the stochastic transition matrices of the models. Both issues are addressed in this volume. The explicit computation of transition probabilities, in particular, allows for the application of the theory of Markov chain aggregation in order to reduce the state space of the model.

2.2.1 A Very Short Introduction to the Markov Chain Setting

For the purposes of this book, it is not necessary to provide an extensive overview of Markov chain theory. It is more convenient here to introduce the general idea for using Markov chains for the representation of ABMs and introduce the analysis tools of Markov chain theory when we apply them to the models. In Chap. 4, for instance, we will analyze the voter model on the complete graph which gives rise to an absorbing birth-death process known as Moran process (Moran 1958). The standard tools for the analysis of absorbing chains are introduced and applied there. In the same way Chap. 6 can be consulted for the analysis of regular Markov chains. In the applications of Markov chain tools presented throughout this book we mainly follow Kemeny and Snell (1976), Behrends (2000), and Levin et al. (2009). Many other volumes (introductory and advanced) are available.

Here we concentrate on ABMs with a finite number of agents that are characterized by a finite set of discrete attributes. This means that the state space of the system—that is, the set of all possible system configurations—is also finite.[1] It will be denoted as Σ in the sequel. Furthermore, AB simulation models usually implement time-discrete processes and due to these ingredients taken together we concentrate on finite-state, discrete-time processes.

A Markov chain is a stochastic process in which the probability to observe a state $\mathbf{y}$ at time $t + 1$ is completely determined by the preceding state $\mathbf{x}$ at time t. It is common to express this in form of a transition probability matrix $\hat{P} : \Sigma \to \Sigma$ that contains the transition probabilities for all pairs of states $\mathbf{x}, \mathbf{y} \in \Sigma$. Then, considering a initial distribution $\hat{\pi}(0)$ that assigns an initial probability to all the possible system states, the time evolution is given by the repeated application of the transition matrix $\hat{\pi}(t) = \hat{\pi}(0)\hat{P}^t$ where $\hat{\pi}(t)$ now contains the probability for the system states at time t.

---

[1] Notice, that this excludes a series of models (e.g. continuous opinion dynamics Deffuant et al. 2001; Hegselmann and Krause 2002) that operate with agents characterized by a continuous variable.

Throughout this book, we will mainly be confronted with two different classes of Markov chains, namely, absorbing and regular chains. This first ones are characterized by the fact that there are certain states $\mathbf{x}$ in the system with no outgoing probabilities, meaning that the system will remain in $\mathbf{x}$ once it has entered it. In other words, $\hat{P}(\mathbf{x}, \mathbf{x}) = 1$ and the process is said to converge to the absorbing state $\mathbf{x}$. For this reason, questions concerning convergence times and the number of times the non-absorbing, transient states are visited before convergence are among the most interesting. As already mentioned, the tools to address those questions are introduced in Chap. 4.

Regular chains, to the contrary, are characterized by the fact that there is a certain time t at which the matrix $\hat{P}^t$ has only positive elements. This obviously excludes absorbing states as the respective outgoing transition probabilities for these states will always remain zero. It basically means that in regular chains every state can be reached from every other state in the course of the process. Moreover, the powers of the transition probability matrix approach a limiting matrix for $t \to \infty$ in which all rows are the same probability vector. Therefore, independent of the initial distribution $\hat{\pi}(0)$, a regular chain always converges to a fixed probability vector $\hat{\pi}$ which is called the stationary distribution of the chain. Since the stationary vector $\hat{\pi}$ is constant under further application of the transition matrix, one way to compute this vector is solve the eigenvalue problem $\hat{\pi}\hat{P} = \hat{\pi}$ (for the eigenvalue 1). Chapter 6 will deal with an ABM that gives rise to regular chains.

When simulating an ABM, one usually initializes the system with particular (often random) initial assignments of the agent attribute corresponding to one specific system configuration $\mathbf{x}$. The initial distribution corresponds in this case to a vector that contains zero everywhere except for the element representing $\mathbf{x}$ where it is one (i.e., $\hat{\pi}_{\mathbf{x}}(0) = 1$ and $\hat{\pi}_{\mathbf{y}}(0) = 0$, $\forall \mathbf{y} \neq \mathbf{x}$). However, in order to understand the dynamics of a model, a series of numerical experiments is usually performed each with a different initial condition. This can be accounted for by setting $\hat{\pi}$ accordingly. One of the strength of using Markov chains is then that the statistics one derives from the analysis accounts for the statistics that would be observed for infinitely many model realizations.

2.3 Lumpability and State Space Aggregation

The state space of a Markov chain derived by considering as states all possible system configurations is far too big to directly use the respective transition matrix $\hat{P}$ for exact numerical computations. As an example, consider a model with binary agent attributes such as the VM. A system of N agents will lead to a Markov chain of size 2^N which for our introductory example of only 20 agents (Fig. 1.1) leads to a chain with more than a million states. In order to use the Markov chain machinery for AB systems, the system size has to be reduced in some way.

2.3.1   Strong Lumpability

This brings lumpability into play as a way to combine and aggregate the states of
a Markov chain so that the process at the aggregate level is still a Markov chain.
Consider that the state space of a Markov chain is Σ and the transition probabilities
between all pairs of states in Σ are given by the $|\Sigma| \times |\Sigma|$ transition matrix $\hat{P}$.
Throughout this work, the chain $(\Sigma, \hat{P})$ will be called micro chain and, respectively,
the states in Σ micro states. Now assume that $\mathbf{X} = (X_0, X_1, \ldots, X_n)$ is a partition
of Σ where each X_k contains a set of micro states in Σ, such the X_k are disjoint
$(X_k \cap X_s = \emptyset$ for any pair of aggregate sets) and for the union of all sets $\bigcup_{i=0}^{n} X_i =$
Σ. Such a situation naturally arises if the process is observed not at the micro level
of Σ, but rather in terms of a measure on Σ, $\phi : \Sigma \to \{0, 1, \ldots, n\}$, by which
all states in Σ that give rise to the same measurement are mapped into the same
aggregate set X_k (also referred to as macro states). An important question that arises
in such a setting is whether the new aggregate process on $\mathbf{X}$ is still a Markov chain
or not. This is what lumpability is about. The lumpability theory adopted for the
purposes of this thesis is largely based on Kemeny and Snell (1976), which is, to the
authors knowledge, the first textbook in which the strong as well as the weak form
of lumpability are discussed with some detail. Notice that there are some other early
and seminal works on lumpability, such as Burke and Rosenblatt (1958), Rosenblatt
(1959), and Rogers and Pitman (1981).

To illustrate the concept of *strong lumpability*, let us use the Land of Oz
example repeatedly considered in Kemeny and Snell (1976) (see pages 29/30 for
the introduction of the example and page 125 for the lumpability example). There, a
three-state Markov chain is formed which approximates how the whether develops
from 1 day to the other. There is rain (R), nice whether (N) and snow (S) and the
transition rates are given by

$$\hat{P} = \begin{matrix} R \\ N \\ S \end{matrix} \begin{pmatrix} 1/2 & 1/4 & 1/4 \\ 1/2 & 0 & 1/2 \\ 1/4 & 1/4 & 1/2 \end{pmatrix}. \tag{2.1}$$

Therefore, a nice day is never followed by a nice day, but there is an equal chance
to have rain or snow. For a rainy day as well as for a day with snow, on the contrary,
there is a chance of $1/2$ that the whether remains as it is for the next day, and the
remaining options are equally likely with probability $1/4$. From this assignment
of probabilities, we can already see that the behavior for rain (R) and snow (S) is
actually equal and therefore we may combine the two states into a "macro" state
called "bad whether" $(B = \{R, S\})$. Hence, the states space is partitioned into two
sets: N on the one hand and $B = \{R, S\}$ on the other. Now, as the probability that
nice whether follows is equal for R and S the transition matrix of the new chain is
uniquely defined by:

$$P = \begin{matrix} N \\ B \end{matrix} \begin{pmatrix} 0 & 1 \\ 1/4 & 3/4 \end{pmatrix}. \tag{2.2}$$

It is the equality of conjoint transition rates from the states that shall be combined to all the other partitions ($\hat{P}(R, N) = \hat{P}(S, N) = 1/4$ in this simple example) on which the condition for lumpability is based.

More precisely, if the probability of moving from a micro state $\mathbf{x} \in X_k$ to a macro state X_l is equal for all micro states in X_k, then all the information about the history which led to a particular state in X_k is actually irrelevant, because from the macro perspective the future evolution is equivalent for any state in X_k. This leads to a condition on the transition matrix $\hat{P}$, namely, $\sum_{\mathbf{y} \in X_l} \hat{P}(\mathbf{x} \in X_k, \mathbf{y} \in X_l)$ must be equal for all $\mathbf{x} \in X_k$. For a process to be lumpable with respect to a partition $\mathbf{X}$, it is sufficient and necessary if this is true for any pair of sets X_k, X_l of the partition. The respective theorem is presented in Kemeny and Snell (1976, Theorem 6.3.2) and we will come back to it with more detail and a focus on an application to ABMs in Sect. 3.3.3 (next chapter).

If the chain along with the desired state space partition is given, the application of the conditions provided in Kemeny and Snell (1976, Theorem 6.3.2) (as well as the subsequent matrix conditions) is relatively simple. However, if only the chain is given, it may be a real challenge to find partitions with respect to which the process is lumpable, not least due to the combinatorial explosion of the number of possible partitions. In this context, some algorithms have been presented for the task to find the optimal or coarsest partition (Buchholz 2000; Derisavi et al. 2003). Other authors have addressed these issues by studying the spectral properties of lumpable chains and have proposed algorithms based on that (Barr and Thomas 1977; Meila and Shi 2001; Takacs 2006; Jacobi 2008; Filliger and Hongler 2008; Görnerup and Jacobi 2010).

Another approach in which aggregate Markov chain descriptions are derived on the basis of model specifications that include the hierarchical and symmetric composition of sub-models has been followed by Buchholz (1995) and is also advised in the context of interactive Markov chains by Hermanns (1999) and Hermanns and Katoen (2010). Namely for systems that "include a large number of identical and symmetric components" (Buchholz 1995, pp. 93/94), a reduced Markov chain description "resulting from exact lumping" (Buchholz 1995, p. 94) is constructed directly during the modeling process. This avoids time-consuming (up to unfeasibility) computations on the huge transition matrices that the model would give rise to without the reduction. In this work, we formulate explicitly the complete microscopic system—containing all symmetries that come by the ABM at question—and lumpability arguments are based on that description (Sects. 3.2 and 3.3, next chapter). However, one of the main messages of this work concerns the translation of model symmetries into regularities on the associated micro chain which then enable lumpability. Especially Chap. 5, in which aggregate descriptions are derived starting from the symmetries of the agent network, is clearly related to the hierarchical approach due to Buchholz (1995) and the idea of symmetric composition in Hermanns (1999).

2.3.2 Weak Lumpability

This thesis mostly applies the strong version of lumpability described above in order to achieve a Markovian aggregation for ABMs. However, it is important to note that there is a weaker version of lumpability often referred to as *weak lumpability* which will play some role in the seventh chapter. While in the case of strong lumpability the projected process on $\mathbf{X} = \{X_0, X_1, \ldots\}$ is a Markov chain for any (initial) distribution, the weaker form of lumpability makes statements about the possibility to obtain a Markovian process at the aggregate level only for particular initial vectors.

For a description of the intuition behind weak lumpability the reader is encouraged to have a look to Kemeny and Snell (1976, Sect. 6.4., and pages 132/133 in particular) who themselves refer to Burke and Rosenblatt (1958) for some of their results. The main idea resides in the following possibility:

> Assume that no matter what the past information is, we always end up with the same assignment of probabilities for being in each of the states in $[X_k]$. Then again the past can have no influence on our predictions. (Kemeny and Snell 1976, p. 133)

A necessary and sufficient (though not always practical) condition (Kemeny and Snell 1976, Theorem 6.4.1) is also provided, but the necessity and sufficiency of conditions for weak lumpability have also been subject of further discussion, see Abdel-Moneim and Leysieffer (1982), Rubino and Sericola (1989), and Peng (1996).

On of the most important observations concerns the fact that if a regular chain is weakly lumpable with respect to a partition $\mathbf{X}$ for some probability vector, then it is weakly lumpable for the stationary vector (the left invariant vector of the transition matrix $\pi P = \pi$). See Kemeny and Snell (1976, Theorem 6.4.3) and also Rubino and Sericola (1989). This may be useful for the decision whether there is one distribution altogether for which a chain is weakly lumpable or not (Kemeny and Snell 1976, Theorem 6.4.4). This result has been extended to absorbing Markov chains by Ledoux et al. (1994). In the absorbing case, the quasi-stationary distribution is shown to play the role of the stationary vector which allows to relate the lumpability problem and existing algorithms for irreducible chains to the absorbing case.

2.3.3 Nearly Lumpable and Non-lumpable Aggregation

It is well known that lumpability (the strong as well as the weak version) is rather an exception than the rule (Chazottes and Ugalde 2003; Gurvits and Ledoux 2005). Some form of aggregation, state space reduction, or macroscopic observation, however, is omnipresent in the analysis of complex systems and their dynamics. The question that then arises concerns the extend to which an aggregate process still informs us about the real microscopic model behavior.

There are some works that discuss these issues for the cases that the aggregation satisfies different types of lumpability. Namely, Schweitzer (1984), Sumita and Rieders (1989), and Buchholz (1994) show that important stationary and transient measures are preserved by the lump. However, the direct derivation of stationary and transient properties of the original chain only by knowledge of the aggregated chain is possible only for a special case of weak lumpability referred to as *exact lumpability* (Buchholz 1994, Theorem 3, Theorem 6). Buchholz (1994) also states that for any micro process and any partition it is possible to construct an aggregation that preserves the stationary measure. However, for the construction of this so-called *ideal aggregate* the stationary state of the original micro system has to be known. Though all lumpable aggregation are also ideal, the converse is not true and Buchholz (1994, p. 6) states:

> In all cases considered here, no information about the transient behavior can be gained from the ideal aggregate.

In Chap. 7 of this work, we will construct an ideal non-lumpable aggregate for the contrarian VM on networks. While this book does not go much further in analyzing the relation between that ideal aggregate and the micro process, it does present an analytical example in which these questions can be addressed.

A second important contribution due to Schweitzer (1984) and Buchholz (1994) is an operational concept of *near lumpability*. The main idea is that a nearly lumpable transition matrix $\hat{P}$ can be represented as $\hat{P} = \hat{A} + \epsilon \hat{B}$ where $\hat{A}$ is lumpable and ϵ is a sufficiently small constant used in analogy to its use in perturbation theory. Buchholz (1994) constructs bounding matrices for the transition probabilities that can be used to compute bounds for the stationary and transient quantities of the aggregated process. The computation of bounds in Buchholz (1994) is in part based on the work of Courtois and Semal (1984). See also Franceschinis and Muntz (1994) and Dayar and Stewart (1997) for other concepts of nearly- or quasi-lumpability.

2.3.4 Aggregation in Dynamical Systems

Finally, to complete this section, we should notice that aggregation and state space decomposition is a wide field which has been vividly discussed across different disciplines, during quite some time. In philosophy, it relates strongly to the more general discussions about the decomposability of a complex system (Simon 1962) and from there to emergence (Wimsatt 1986; Auger and Poggiale 1998) and even further to the possible limitations of an reductionist account of complex systems (Wimsatt 2006a). In economics, where much theory is in fact developed around aggregate measures, techniques for the aggregation of variables in dynamical systems have been developed (e.g., Theil 1965; Simon and Ando 1961; Ando and Fisher 1963) as an operationalization "decomposability" and "nearly-decomposability" of a complex system mentioned above (Simon 1962). These techniques have been transferred to theoretical biology, ecological modeling and population dynamics in particular, by Iwasa et al. (1987) in which conditions for exact aggregation in non-linear dynamical systems are given and Iwasa et al. (1989) which deals with approximate aggregations. The fact that the explicit consideration of more and more factors is a tendency in modern model development, has led to a renewed interest in aggregation techniques not only in Markov chains but also in the context of dynamical systems (see Auger et al. 2008 for a review of aggregation methods with application to population dynamics).

It is clear that aggregation techniques are actually relevant to all models which involve a large number of variables (or agents), in order to derive reduced model descriptions that might be amenable to analytical strategies. Markov chains and dynamical systems are probably the two most important mathematical formalisms to represent complex and high-dimensional systems that evolve in time. In this context, it is very interesting that methods for aggregation of variables in linear dynamical systems and lumpability in Markov chains can be based on the same principles, a fact that has recently been exploited in Jacobi and Görnerup (2009) and Görnerup and Jacobi (2010).

2.4 The Information-Theoretic Perspective

A useful complementary view on lumpability and state space aggregation more generally is provided by a series of information-theoretic approaches that are recently developed in the context of multi-level dynamical systems (Shalizi and Moore 2003; Görnerup and Jacobi 2008, 2010; Jacobi and Görnerup 2009; Pfante et al. 2014a,b). Albeit being applied to dynamical systems more generally, the setting is strongly related to the questions of lumpability in Markov chains. Consider a Markov chain $(\Sigma, \hat{P})$ with state space Σ and a transition matrix $\hat{P}$ and an operator $\phi : \Sigma \to \mathbf{X}$ that projects the system onto a higher-level coarse-graining $\mathbf{X}$ of Σ inducing a dynamical process on $\mathbf{X}$. The question of lumpability is basically whether the induced process on the $\mathbf{X}$-level is still Markovian.

In the previous section, we have somehow considered that the partition $\mathbf{X}$ is already defined. This is reasonable in many cases, for instance, in most AB studies where the system property one wishes to analyze defines a projection (see Chap. 3). However, in multi-level systems more generally, the state space partition $\mathbf{X}$ might not be known beforehand. This leads to questions of level-identification where one has to find projection operators (and consequently partitions) that lead to a "closed" description (at least approximately), in the sense that the system can be modeled by the state variables of this level. Information-theoretic measures can be used in order to quantify "closedness", or, to be precise, deviations from it. Here we shall mention three of these measures:

Markovianity Shalizi and Moore (2003) emphasize the particular role of Markovianity in the definition or identification of macroscopic observables. Based on that, Görnerup and Jacobi (2008) propose a Markovianity measure following the idea that an higher level is closed if the dynamic $P : \mathbf{X} \to \mathbf{X}$ induced at this level is Markovian. The decision whether the macro process (obtained by a certain projection) is Markovian or not is based on the mutual information between the past $(\ldots, X_{t-2}, X_{t-1})$ and the future $(X_{t+1}, X_{t+2}, \ldots)$ with respect to the present (X_t). If the expected mutual information between past and future is zero, looking further back into the past does not provide any new information about the future evolution, that is, the future depends only on the present value X_t and the sequence induced at the macro level is a Markov process. In other words, the conditional past-future mutual information $I(X_{t+1}; X_{-\infty}^{t-1}|X_t)$ vanishes. Noteworthy, they show that their Markovianity measure can be expressed in terms of the slope of block entropies which bears a relation to process reconstruction in turbulence and finance (Chazottes et al. 1998; Vilela Mendes et al. 2002).

Informational Closure According to this measure, introduced in Pfante et al. (2014a), a level is informational closed if the knowledge of the micro-level state $\mathbf{x}_t$ at time t does not allow for better predictions of the macro level X_{t+1} than the knowledge of the preceding macro state X_t. This can be written as the conditional mutual information $I(X_{t+1}; \mathbf{x}_t|X_t)$ which quantifies the information flow from the original to the higher level. In other words, this measure quantifies micro-level information that a higher-level description does not account for and consequently a level is closed if $I(X_{t+1}; \mathbf{x}_t|X_t)$ vanishes. As shown in Pfante et al. (2014a), $I(X_{t+1}; X_{-\infty}^{t-1}|X_t) \leq I(X_{t+1}; \mathbf{x}_t|X_t)$ so that vanishing information flow from micro to macro implies Markovianity. Moreover, in most situations information flow can distinguish between the strong and the weak form of lumpability as it vanishes for the former but not for the latter.

Predictive Efficiency The intuition behind predictive efficiency, introduced in Shalizi (2001) with important predecessors in Grassberger (1986), Lindgren and Nordahl (1988), and Crutchfield and Young (1989) (among others), is that a coarse-grained description with state space $\mathbf{X}$ can be considered as a level if it is informative for the dynamics at this level while, at the same time, being not too complex. Shalizi (2001) introduces the notion as the ratio between excess entropy and statistical complexity and uses it to define emergent processes. Based on this, two variants of predictive efficiency are introduced in Pfante et al. (2014b): first, the ratio $I(X_{t+1}; X_t)/H(X_t)$ between one-step mutual (prediction) information and the entropy of the description; second, the variational $I(X_{t+1}; X_t) - \beta H(X_t)$ which relates the measure to the information bottleneck method (Tishby et al. 1999).

We will come back to these measures in the seventh chapter where we study a non-absorbing variant of the VM on a two-community graph. The projection of the micro dynamics of this model onto the macroscopic level is not lumpable which means that memory effects are introduced in the transition from the micro to the macro level. For the special two-community case we are able to compute Markovianity and informational closure explicitly.

The information-theoretic setting described in this section is also related to the framework of computational mechanics (Crutchfield and Young 1989; Shalizi and Crutchfield 2001; James et al. 2011, and references therein). The main idea in computational mechanics is to group histories which give rise to the same conditional probability distribution over futures into equivalence classes—so-called causal states—and to construct in this way a minimal causal model—called ϵ-machines—for the prediction of the process at question. The reader may be referred to Shalizi and Crutchfield (2001) for an overview and several interesting theoretical results in computational mechanics. The applicability of these measures to AB and related computational models is limited by their computational complexity (cf. Görnerup and Jacobi 2008, p. 13). The fact that, even in very simple ABMs, the state space of the process to be handled becomes very large challenges these approaches in two ways. The first one concerns the "combinatorial explosion" (Görnerup and Jacobi 2008, p. 11) of the number of possible partitions, which is a general difficulty for level identification where the partition is not given a priori. Secondly, the larger the alphabet, the more data must be generated and evaluated in order to obtain a workable approximation of the joined probability distribution of sequence blocks (cf. Shalizi and Crutchfield 2001, Sect. VII.B/C). One way to deal with this problem is to restrict to block size to one, as in Shalizi et al. (2004), which is actually exact if the original process is a Markov chain. Still, in this case, the number of states is huge and the estimation of the conditional probabilities (on the basis of which equivalence classes are constructed) requires a lot of simulation data.

2.5 Motivation: Towards a Markov Chain Theory of Aggregation for Agent-Based Models

2.5.1 Bridging a Gap

Though it has often been recognized that ABMs may be conceived as (stochastic) dynamical systems or Markov chains (Epstein and Axtell 1996; Laubenbacher et al. 2009; Izquierdo et al. 2009; Page 2012), the afore mentioned aggregation techniques developed for these systems have not yet been applied to ABMs. One of the reasons for this is that an explicit formulation of the micro process in terms of dynamical systems or Markov chains has been accomplished only in an abstract (Laubenbacher et al. 2009; Page 2012) or approximate (Izquierdo et al. 2009) way. The explicit formalization of the micro process as a Markov chain—the reasoning presented in this book will be started with it (Sect. 3.2)—enables the application of the Markov chain theory of aggregation—that is, lumpability—to ABMs.

The need for a mathematical framework that links the micro and the macro level has, of course, been noted earlier. For instance:

> Of course, microscopic and macroscopic theories are related, and understanding the connection between the two, e.g., through simulation or by deriving the latter from the former, is an important goal of any complex systems research. (Lerman 2001, p. 225)

Also the general possibility of applying mathematical aggregation techniques (Page 2012) and complexity reduction by symmetry exploitation (Laubenbacher et al. 2009) has been noted, namely, in the context of dynamical systems and partly based on earlier work by Iwasa et al. (1987) in population ecology. However, a sophisticated and practicable mathematical framework for linking between micro and macro level processes in an AB system does not yet exist. This work is a first step to bridge this gap.

2.5.2 The Micro-Macro Link

The relation between the microscopic and the macroscopic has since long been subject for controversy. In sociology, it is manifest in the dichotomy of methodological individualism and structural functionalism. A good overview over the historical development of micro-macro debates from philosophy to social theory is provided in the introductory chapter (Alexander and Giesen 1987) of a volume headed "The Micro-Macro Link" (Alexander et al. 1987).

"The Micro-Macro Link" is a collection of essays by very influential social theorists in the micro as well as in the macro tradition about ways to overcome the micro-macro divide and link between the different levels of analysis. A synthetic formulation embracing the different levels from individual action to social order and back requires on the one hand a link from the micro to the macro pointing at questions related to various (from weaker to stronger) forms of emergence (Brodbeck 1968; Giesen 1987), aggregation and equilibrium (Coleman 1987). On the other, it should also include concepts for the retro-action of the macro on the micro level, such as internalization (Parsons 1954) or constraints on and the environment of individual actions (Alexander 1987). One of the first acknowledged synthetic formulations of this linkage between micro and macro in sociology studies is from Max Weber (1978) from where we quote the following basic observation:

> within the realm of social action, certain empirical uniformities can be observed, that is, courses of action that are repeated by the actor or (simultaneously) occur among numerous actors (Weber 1978, p. 29)

We shall see how a stylized version of this belief is incorporated in our study when passing from micro to macro dynamics.

AB simulation is sometimes considered as a methodology to provide a "theoretical bridge" (Macy and Willer 2002, p. 148) between micro and macro theories (see also Saam 1999; Squazzoni 2008). Even if most of the models (especially the early ones) are actually a straight implementation of the individualistic program, there are some attempts to include into the model agents with some socio-cognitive abilities (see Squazzoni 2008, pp. 14–16) capable of the perception and internalization of the macro sphere. Also the experimentation with different interaction topologies can actually be seen as an attempt to understand the influence of social structure (macro) on the emergence of collective order (macro) transmitted through the level of individual interaction (micro).

Clearly, this book is not about social theory. It is about a mathematical technique to link micro dynamics to macro dynamics in models that may be designed on the basis of sociological theorizing. To my opinion, a well posed mathematical basis for these models may help the understanding of many of their observed properties, and it also provides a new perspective on aggregation and emergence and on how they are related. Linking the micro-description of an ABM to a macro-description in the form of a Markov chain provides information about the transition from the interaction of individual actors to the complex macroscopic behaviors observed in social systems. In particular, well-known conditions for lumpability (Sect. 2.3.1) make it possible to decide whether the macro model is still Markov. Conversely, this setting can also provide a suitable framework to understand the emergence of long range memory effects and patterns of spatial organization (Chap. 7).

2.5.3 Computational Emergence and Aggregativity

ABMs and other related computational tools (such as CA) play an increasingly important role also in the contemporary philosophical discussions of emergence. Some philosophers (e.g., Bedau 1997, 2003; Huneman and Humphreys 2008; Humphreys 2008) advocate a position which makes use of computational models as a playground to address fundamental questions of emergence (see Symons 2008 for a critical consideration). Questions about the relation of these artificial model environments to real phenomena are not ignored, but considered as an independent issue which is actually part of another debate. The field of computational emergence aims to establish "a close link between the concept of emergence and computation or computer simulations, which can perhaps be captured by the idea that an emergent phenomenon is one that arises from a computationally incompressible process" (Huneman and Humphreys 2008, pp. 425/426). The framework presented here provides explicit knowledge about the (in)compressibility of computational models and the dynamical processes which these models give rise to.

While scientists use the term "emergence" relatively freely, the philosophical literature differentiates more carefully between different forms of emergence (onto-logical versus epistemological, strong versus weak, synchronic versus diachronic emergence) and the existence of some of these forms (ontological emergence in particular) is in fact highly controversial. In the context of computational models, emergence is often paraphrased by "the whole is more than the sum of its parts" and an emergent property can be a certain macro-level pattern that could not be expected (and not predicted!) by looking at the micro level rules only. Along this lines, a well-known and explicitly computational account of weak emergence that fits the use of the term in complexity science has been offered by Bedau (1997, 2003):

> The behavior of weakly emergent systems cannot be determined by any computation that is essentially simpler than the intrinsic natural computational process by which the system's behavior is generated. (Bedau 2003, p. 18)

Bedau (2003) uses CAs to illustrate these ideas and makes explicit reference to simulations: according to him a system property is emergent if it can be derived "only by simulation" (Bedau 2003, p. 15).

An alternative position on emergence has been advocated by Wimsatt (1986) even before computer simulations became widespread. Wimsatt (1986) starts out from analyzing the conditions for a system property to be a mere aggregate of the properties of the parts of which the system is composed (see also Wimsatt 2000, 2006a,b). Accordingly, a property of a system is called emergent if it does not satisfy these condition for aggregativity. In this way, Wimsatt is able to give a rather straightforward meaning to the dictum "a complex system is more than the sum of its parts" by relating emergence to the lack of aggregativity. What makes Wimsatt's position particularly interesting for this work is not only that relation between aggregation and emergence, but also the observation expressed by the following statements:

> [I]t is better to talk about properties of systems and their parts, and to analyze aggregativity as a kind of relation between these properties. (Wimsatt 1986, p. 260)

> Aggregativity and emergence concern the relationship between a property of a system under study and properties of its parts. (Wimsatt 2006a, p. 675)

The reason for which it is better to focus on properties, or rather to be explicit on that point, is that a system might be aggregative for one but emergent for another property. Just as a Markov chain might be lumpable with respect to one but non-lumpable with respect to another partition!

References

Abdel-Moneim, A. M., & Leysieffer, F. W. (1982). Weak lumpability in finite Markov chains. *Journal of Applied Probability, 19*(3), 685–691.

Alexander, J. C. (1987). Action and its environments. In J. C. Alexander, B. Giesen, R. Münch, & N. J. Smelser (Eds.), *The micro–macro link*. Berkeley: University of California Press.

Alexander, J. C., & Giesen, B. (1987). From reduction to linkage: The long view of the micro–macro link. In J. C. Alexander, B. Giesen, R. Münch & N. J. Smelser (Eds.), *The micro–macro link*. Berkeley: University of California Press.

Alexander, J. C., Giesen, B., Münch, R., & Smelser, N. J. (Eds.). (1987). *The micro–macro link*. Berkeley: University of California Press.

Ando, A., & Fisher, F. M. (1963). Near-decomposability, partition and aggregation, and the relevance of stability discussions. *International Economic Review, 4*(1), 53–67.

Auger, P., de La Parra, R. B., Poggiale, J.-C., Sánchez, E., & Sanz, L. (2008). Aggregation methods in dynamical systems and applications in population and community dynamics. *Physics of Life Reviews, 5*(2), 79–105.

Auger, P., & Poggiale, J.-C. (1998). Aggregation and emergence in systems of ordinary differential equations. *Mathematical and Computer Modelling, 27*(4), 1–21.

Axelrod, R. (1997). The dissemination of culture: A model with local convergence and global polarization. *The Journal of Conflict Resolution, 41*(2), 203–226.

Axelrod, R. (2003). Advancing the art of simulation in the social sciences. *Japanese Journal for Management Information System, Special Issue on Agent-Based Modeling, 12*(3), 1–19. Updated article originally published in simulating social phenomena (pp. 21–40). Berlin: Springer (1997).

Axtell, R., Axelrod, R., Epstein, J. M., & Cohen, M. D. (1996). Aligning simulation models: A case study and results. *Computational and Mathematical Organization Theory, 1*(2), 123–141.

Banisch, S., Araujo, T., & Louca, J. (2010). Opinion dynamics and communication networks. *Advances in Complex Systems, 13*(1), 95–111. ePrint: arxiv.org/abs/0904.2956.

Barr, D. R., & Thomas, M. U. (1977). An eigenvector condition for Markov chain lumpability. *Operations Research, 25*(6), 1028–1031.

Bedau, M. A. (1997). Weak emergence. *Philosophical Perspectives, 11*, 375–399.

Bedau, M. A. (2003). Downward causation and the autonomy of weak emergence. *Principia Revista Internacional de Epistemologica, 6*(1), 5–50.

Behrends, E. (2000). *Introduction to Markov chains with special emphasis on rapid mixing* (Vol. 228). Braunschweig/Wiesbaden: Vieweg Springer.

Berlekamp, E., Conway, J., & Guy, R. (1982). *Winning ways for your mathematical plays* (Vol. 2). London: Academic.

Bornholdt, S. (2001). Expectation bubbles in a spin model of markets: Intermittency from frustration across scales. *International Journal of Modern Physics C, 12*, 667–674.

Brodbeck, M. (1968). Methodological individualism - definitions and reduction. In M. Brodbeck (Ed.), *Readings in the philosophy of the social sciences* (pp. 280–309). London: Collier-Macmillan.

Buchholz, P. (1994). Exact and ordinary lumpability in finite Markov chains. *Journal of Applied Probability, 31*(1), 59–75.

Buchholz, P. (1995). Hierarchical Markovian models: Symmetries and reduction. *Performance Evaluation, 22*(1), 93–110. 6th International Conference on Modelling Techniques and Tools for Computer Performance Evaluation.

Buchholz, P. (2000). Efficient computation of equivalent and reduced representations for stochastic automata. *Computer Systems Science & Engineering, 15*(2), 93–103.

Burke, C. J., & Rosenblatt, M. (1958). A Markovian function of a Markov chain. *The Annals of Mathematical Statistics, 29*(4), 1112–1122.

Castellano, C., Fortunato, S., & Loreto, V. (2009). Statistical physics of social dynamics. *Reviews of Modern Physics, 81*(2), 591–646.

Castellano, C., Marsili, M., & Vespignani, A. (2000). Nonequilibrium phase transition in a model for social influence. *Physical Review Letters, 85*(16), 3536–3539.

Chazottes, J.-R., Floriani, E., & Lima, R. (1998). Relative entropy and identification of Gibbs measures in dynamical systems. *Journal of Statistical Physics, 90*(3–4), 697–725.

Chazottes, J.-R., & Ugalde, E. (2003). Projection of Markov measures may be Gibbsian. *Journal of Statistical Physics, 111*(5/6), 1245–1272.

Claussen, J. C., & Traulsen, A. (2005). Non-Gaussian fluctuations arising from finite populations: Exact results for the evolutionary Moran process. *Physical Review E, 71*(2), 025101.

Coleman, J. S. (1987). Microfoundations and macrosocial behavior. In J. C. Alexander, B. Giesen, R. Münch & N. J. Smelser (Eds.), *The micro–macro link*. Berkeley: University of California Press.

Cont, R., & Bouchaud, J.-P. (2000). Herd behavior and aggregate fluctuations in financial markets. *Macroeconomic Dynamics, 4*(2), 170–196.

Corcuera, J. M., Nualart, D., & Schoutens, W. (2005). Completion of a Lévy market by power-jump assets. *Finance and Stochastics, 9*(1), 109–127.

Courtois, P.-J., & Semal, P. (1984). Bounds for the positive eigenvectors of nonnegative matrices and for their approximations by decomposition. *Journal of the ACM, 31*(4), 804–825.

Cox, J. T. (1989). Coalescing random walks and voter model consensus times on the torus in Zd. *The Annals of Probability, 17*(4), 1333–1366.

Crutchfield, J. P., & Young, K. (1989). Inferring statistical complexity. *Physical Review Letters*, *63*(2), 105–108.

David, N., Sichman, J. S., & Coelho, H. (2005). The logic of the method of agent-based simulation in the social sciences: Empirical and intentional adequacy of computer programs. *Journal of Artificial Societies and Social Simulation, 8*(4), 2.

Dayar, T., & Stewart, W. J. (1997). Quasi lumpability, lower-bounding coupling matrices, and nearly completely decomposable Markov chains. *SIAM Journal on Matrix Analysis and Applications, 18*(2), 482–498.

De Oliveira, M., Mendes, J., & Santos, M. (1993). Nonequilibrium spin models with Ising universal behaviour. *Journal of Physics A: Mathematical and General, 26*(10), 2317.

Deffuant, G., Neau, D., Amblard, F., & Weisbuch, G. (2001). Mixing beliefs among interacting agents. *Advances in Complex Systems, 3*, 87–98.

Derisavi, S., Hermanns, H., & Sanders, W. H. (2003). Optimal state-space lumping in Markov chains. *Information Processing Letters, 87*(6), 309–315.

Epstein, J. M., & Axtell, R. (1996). *Growing artificial societies: Social science from the bottom up*. Washington, DC: The Brookings Institution.

Filliger, R., & Hongler, M. O. (2008). Lumping complex networks. In *Lectures and Gallery of Madeira Math Encounters XXXV*. http://ccm.uma.pt/mme35/.

Fisher, R. A. (1930). *The genetical theory of natural selection: a complete variorum edition*. Oxford: Oxford University Press.

Frachebourg, L., & Krapivsky, P. L. (1996). Exact results for kinetics of catalytic reactions. *Physical Review E, 53*(4), R3009–R3012.

Franceschinis, G., & Muntz, R. R. (1994). Bounds for quasi-lumpable Markov chains. *Performance Evaluation, 20*(1–3), 223–243. Performance '93.

Galán, J. M., Izquierdo, L. R., Izquierdo, S. S., Santos, J. I., del Olmo, R., López-Paredes, A., & Edmonds, B. (2009). Errors and artefacts in agent-based modelling. *Journal of Artificial Societies and Social Simulation, 12*(1), 1.

Giesen, B. (1987). Beyond reductionism: Four models relating micro and macro levels. In J. C. Alexander, B. Giesen, R. Münch, & N. J. Smelser (Eds.), *The micro-macro link* (Chapter 15). Berkeley: University of California Press.

Gleeson, J. P. (2011). High-accuracy approximation of binary-state dynamics on networks. *Physical Review Letters, 107*, 068701.

Gleeson, J. P. (2013). Binary-state dynamics on complex networks: Pair approximation and beyond. *Physical Review X, 3*, 021004.

Görnerup, O., & Jacobi, M. N. (2008). A method for inferring hierarchical dynamics in stochastic processes. *Advances in Complex Systems, 11*(1), 1–16.

Görnerup, O., & Jacobi, M. N. (2010). A method for finding aggregated representations of linear dynamical systems. *Advances in Complex Systems, 13*(02), 199–215.

Grassberger, P. (1986). Toward a quantitative theory of self-generated complexity. *International Journal of Theoretical Physics, 25*(9), 907–938.

Grauwin, S., Goffette-Nagot, F., & Jensen, P. (2010). *Dynamic models of residential segregation: An analytical solution*. Working Papers 1017, Groupe d'Analyse et de Théorie Economique (GATE), Centre national de la recherche scientifique (CNRS), Université Lyon 2, Ecole Normale Supérieure.

Grimm, V., Berger, U., Bastiansen, F., Eliassen, S., Ginot, V., Giske, J., et al. (2006). A standard protocol for describing individual-based and agent-based models. *Ecological Modelling, 198*, 115–126.

Gurvits, L., & Ledoux, J. (2005). Markov property for a function of a Markov chain: A linear algebra approach. *Linear Algebra and Its Applications, 404*(0), 85–117.

Hales, D., Rouchier, J., & Edmonds, B. (2003). Model-to-model analysis. *Journal of Artificial Societies and Social Simulation, 6*(4), 10.

Hegselmann, R., & Krause, U. (2002). Opinion dynamics and bounded confidence: Models, analysis and simulation. *Journal of Artificial Societies and Social Simulation, 5*(3), 1.

Hermanns, H. (1999). *Interactive Markov chains*. Ph.D. thesis, Friedrich-Alexander Universität Erlangen Nürnberg.

Hermanns, H., & Katoen, J.-P. (2010). The how and why of interactive Markov chains. In F. Boer, M. Bonsangue, S. Hallerstede, & M. Leuschel (Eds.), *Formal methods for components and objects*. Lecture Notes in Computer Science (Vol. 6286, pp. 311–337). Berlin/Heidelberg: Springer.

Hofbauer, J., & Sigmund, K. (2003). Evolutionary game dynamics. *Bulletin of the American Mathematical Society, 40*(4), 479–519.

Hommes, C. H. (2006). Heterogeneous agent models in economics and finance. In L. Tesfatsion & K. Judd (Eds.), *Handbook of computational economics, volume 2: Agent-based computational economics* (Vol. 2, pp. 1109–1186). Amsterdam: Elsevier.

Humphreys, P. (2008). Synchronic and diachronic emergence. *Minds and Machines, 18*(4), 431–442.

Huneman, P., & Humphreys, P. (2008). Dynamical emergence and computation: An introduction. *Minds and Machines, 18*(4), 425–430.

Iwasa, Y., Andreasen, V., & Levin, S. (1987). Aggregation in model ecosystems. I. Perfect aggregation. *Ecological Modelling, 37*(3), 287–302.

Iwasa, Y., Levin, S. A., & Andreasen, V. (1989). Aggregation in model ecosystems II. Approximate aggregation. *Mathematical Medicine and Biology, 6*(1), 1–23.

Izquierdo, L. R., Izquierdo, S. S., Galán, J. M., & Santos, J. I. (2009). Techniques to understand computer simulations: Markov chain analysis. *Journal of Artificial Societies and Social Simulation, 12*(1), 6.

Jacobi, M. N. (2008). A dual eigenvector condition for strong lumpability of Markov chains. In *CoRR*. arxiv.org/abs/0710.1986v2.

Jacobi, M. N., & Görnerup, O. (2009). A spectral method for aggregating variables in linear dynamical systems with application to cellular automata renormalization. *Advances in Complex Systems, 12*(02), 131–155.

James, R. G., Ellison, C. J., & Crutchfield, J. P. (2011). Anatomy of a bit: Information in a time series observation. *Chaos, 21*(3), 7109.

Kaizoji, T., Bornholdt, S., & Fujiwara, Y. (2002). Dynamics of price and trading volume in a spin model of stock markets with heterogeneous agents. *Physica A: Statistical Mechanics and its Applications, 316*(1), 441–452.

Kemeny, J. G., & Snell, J. L. (1976). *Finite Markov chains*. New York: Springer.

Kimura, M., & Weiss, G. H. (1964). The stepping stone model of population structure and the decrease of genetic correlation with distance. *Genetics, 49*, 561–576.

Krapivsky, P. L., & Redner, S. (2003). Dynamics of majority rule in two-state interacting spin systems. *Physical Review Letters, 90*(23), 238701.

Krause, S. M., & Bornholdt, S. (2013). Spin models as microfoundation of macroscopic market models. *Physica A: Statistical Mechanics and Its Applications, 392*(18), 4048–4054.

Laubenbacher, R. C., Jarrah, A. S., Mortveit, H. S., & Ravi, S. S. (2009). Agent based modeling, mathematical formalism for. In R. A. Meyers (Ed.), *Encyclopedia of complexity and systems science* (pp. 160–176). Berlin: Springer.

LeBaron, B. (2000). Agent-based computational finance: Suggested readings and early research. *Journal of Economic Dynamics and Control, 24*(5), 679–702. Springer

Ledoux, J., Rubino, G., & Sericola, B. (1994). Exact aggregation of absorbing Markov processes using the quasi-stationary distribution. *Journal of Applied Probability, 31*, 626–634.

Lerman, K. (2001). Design and mathematical analysis of agent-based systems. In J. Rash, W. Truszkowski, M. Hinchey, C. Rouff, & D. Gordon (Eds.), *Formal approaches to agent-based systems*. Lecture Notes in Computer Science (Vol. 1871, pp. 222–234). Berlin/Heidelberg: Springer.

Levin, D. A., Peres, Y., & Wilmer, E. L. (2009). *Markov chains and mixing times*. Providence: American Mathematical Society.

Li, W. H. (Ed.). (1977). *Stochastic models in population genetics*. Stroudsburg: Dowden, Hutchinson and Ross, Inc.

Liberman, E., Hauert, C., & Nowak, M. (2005). Evolutionary dynamics on graphs. *Nature, 433*(7023), 312–316.

Liggett, T. M. (1999). *Stochastic interacting systems: Contact, voter and exclusion processes.* Grundlehren der mathematischen Wissenschaften (Vol. 324). New York: Springer.

Lindgren, K., & Nordahl, M. G. (1988). Complexity measures and cellular automata. *Complex Systems, 2*(4), 409–440.

Macy, M. W., & Willer, R. (2002). From factors to actors: Computational sociology and agent-based modeling. *Annual Review of Sociology, 28*(1), 143–166.

Maruyama, T. (1974). A simple proof that certain quantities are independent of the geographical structure of population. *Theoretical Population Biology, 5*(2), 148–154.

Meila, M., & Shi, J. (2001). A random walks view of spectral segmentation. In *AI and STATISTICS (AISTATS) 2001*.

Moran, P. A. P. (1958). Random processes in genetics. *Proceedings of the Cambridge Philosophical Society, 54*, 60–71.

Moretti, P., Liu, S., Baronchelli, A., & Pastor-Satorras, R. (2012). Heterogenous mean-field analysis of a generalized voter-like model on networks. *The European Physical Journal B, 85*(3), 1–6.

Nielsen, P. H. (2005). Optimal bonus strategies in life insurance: The Markov chain interest rate case. *Scandinavian Actuarial Journal, 2005*(2), 81–102.

Norberg, R. (2006). Dynamic Greeks. *Insurance: Mathematics and Economics, 39*(1), 123–133.

Nowak, M. (2006). *Evolutionary dynamics: Exploring the equations of live.* Cambridge, MA: Harvard University Press.

Page, S. E. (2012). Aggregation in agent-based models of economies. *The Knowledge Engineering Review, 27*(02), 151–162.

Parsons, T. (1954). *Essays in social theory.* New York: Free Press.

Patzelt, F., & Pawelzik, K. (2013). An inherent instability of efficient markets. *Scientific Reports, 3*, 2784.

Peng, N.-F. (1996). On weak lumpability of a finite Markov chain. *Statistics & Probability Letters, 27*(4), 313–318.

Pfante, O., Bertschinger, N., Olbrich, E., Ay, N., & Jost, J. (2014a). Comparison between different methods of level identification. *Advances in Complex Systems, 17*, 1450007.

Pfante, O., Olbrich, E., Bertschinger, N., Ay, N., & Jost, J. (2014b). Closure measures for coarse-graining of the tent map. *Chaos: An Interdisciplinary Journal of Nonlinear Science, 24*(1), 013136.

Pollicott, M., & Weiss, H. (2001). The dynamics of Schelling-type segregation models and a nonlinear graph Laplacian variational problem. *Advances in Applied Mathematics, 27*(1), 17–40.

Preis, T., Moat, H. S., & Stanley, H. E. (2013). Quantifying trading behavior in financial markets using Google trends. *Scientific Reports, 3*, 1684.

Pugliese, E., & Castellano, C. (2009). Heterogeneous pair approximation for voter models on networks. *EPL (Europhysics Letters), 88*(5), 58004.

Reynolds, C. W. (1987). Flocks, herds, and schools: A distributed behavioral model. *ACM Siggraph Computer Graphics 21*(4), 25–34.

Roca, C. P., Cuesta, J. A., & Sánchez, A. (2009). Evolutionary game theory: Temporal and spatial effects beyond replicator dynamics. *Physics of Life Reviews, 6*(4), 208–249.

Rogers, L. C. G., & Pitman, J. W. (1981). Markov functions. *The Annals of Probability, 9*(4), 573–582.

Rosenblatt, M. (1959). Functions of a Markov process that are Markovian. *Journal of Mathematics and Mechanics, 8*(4), 134–145.

Rubino, G., & Sericola, B. (1989). On weak lumpability in Markov chains. *Journal of Applied Probability, 26*(3), 446–457.

Saam, N. (1999). Simulating the micro-macro link: New approaches to an old problem and an application to military coups. *Sociological Methodology, 29*, 43–79.

Schelling, T. (1971). Dynamic models of segregation. *Journal of Mathematical Sociology, 1*(2), 143–186.

Schuster, P., & Sigmund, K. (1983). Replicator dynamics. *Journal of Theoretical Biology, 100*(3), 533–538.

Schweitzer, P. J. (1984). Aggregation methods for large Markov chains. In *Proceedings of the International Workshop on Computer Performance and Reliability* (pp. 275–286). Amsterdam: North-Holland.

Schweitzer, F., & Behera, L. (2009). Nonlinear voter models: The transition from invasion to coexistence. *The European Physical Journal B - Condensed Matter and Complex Systems, 67*(3), 301–318.

Shakarian, P., Roos, P., & A., J. (2012). A review of evolutionary graphs theory with applications to game theory. *Biosystems, 107*, 66–80.

Shalizi, C. R. (2001). *Causal architecture, complexity and self-organization in the time series and cellular automata* (Doctoral dissertation, University of Wisconsin–Madison).

Shalizi, C. R., & Crutchfield, J. P. (2001). Computational mechanics: Pattern and prediction, structure and simplicity. *Journal of Statistical Physics, 104*(3–4), 817–879.

Shalizi, C. R., & Moore, C. (2003). What is a macrostate? Subjective observations and objective dynamics. In *CoRR*. arXiv:cond-mat/0303625.

Shalizi, C. R., Shalizi, K. L., & Haslinger, R. (2004). Quantifying self-organization with optimal predictors. *Physical Review Letters, 93*, 118701.

Simon, H. A. (1962). The architecture of complexity. *Proceedings of the American Philosophical Society, 106*(6), 467–482.

Simon, H. A., & Ando, A. (1961). Aggregation of variables in dynamic systems. *Econometrica: Journal of The Econometric Society, 29*, 111–138.

Slanina, F., & Lavicka, H. (2003). Analytical results for the Sznajd model of opinion formation. *The European Physical Journal B - Condensed Matter and Complex Systems, 35*(2), 279–288.

Slatkin, M. (1981). Fixation probabilities and fixation times in a subdivided population. *Evolution, 35*(3), 477–488.

Smith, J. M. (1982). *Evolution and the theory of games*. Cambridge: Cambridge University Press.

Sood, V., & Redner, S. (2005). Voter model on heterogeneous graphs. *Physical Review Letters, 94*(17), 178701.

Squazzoni, F. (2008). The micro-macro link in social simulation. *Sociologica, 1*, 2.

Steels, L. (1997). Self-organizing vocabularies. In C. Langton & T. Shimohara (Eds.), *Artificial life V: Proceeding of the Fifth International Workshop on the Synthesis and Simulation of Living Systems* (pp. 179–184). Cambridge: MIT.

Sumita, U., & Rieders, M. (1989). Lumpability and time reversibility in the aggregation-disaggregation method for large Markov chains. *Stochastic Models, 5*(1), 63–81.

Symons, J. (2008). Computational models of emergent properties. *Minds and Machines, 18*(4), 475–491.

Takacs, C. (2006). On the fundamental matrix of finite state Markov chains, its eigensystem and its relation to hitting times. *Mathematica Pannonica, 17*(2), 183–193.

Taylor, P. D., & Jonker, L. B. (1978). Evolutionary stable strategies and game dynamics. *Mathematical Biosciences, 40*(1), 145–156.

Tesfatsion, L., & Judd, K. L. (2006). *Handbook of computational economics, volume 2: Agent-based computational economics*. Amsterdam: North-Holland.

Theil, H. (1965). *Linear aggregation of economic relations*. Contributions to economic analysis. Amsterdam: North-Holland.

Tishby, N., Pereira, F. C., & Bialek, W. (1999). The information bottleneck method. In *Proceedings of the 37th Annual Allerton Conference on Communication, Control and Computing* (pp. 368–377).

Traulsen, A., Claussen, J. C., & Hauert, C. (2005). Coevolutionary dynamics: From finite to infinite populations. *Physical Review Letters, 95*, 238701.

Vazquez, F., & Eguiluz, V. M. (2008). Analytical solution of the voter model on uncorrelated networks. *New Journal of Physics, 10*(6), 063011.

Vilela Mendes, R., Lima, R., & Araújo, T. (2002). A process-reconstruction analysis of market fluctuations. *International Journal of Theoretical and Applied Finance, 5*(08), 797–821.

von Neumann, J. (1951). *The general and logical theory of automata* (pp. 1–41). Pasadena, CA: Wiley.

Voorhees, B. (2013). Birth–death fixation probabilities for structured populations. *Proceedings of the Royal Society A: Mathematical, Physical and Engineering Science, 469*(2153), 20248.

Voorhees, B., & Murray, A. (2013). Fixation probabilities for simple digraphs. *Proceedings of the Royal Society A: Mathematical, Physical and Engineering Science, 469*(2154), 20676.

Weber, M. (1978). *Economy and society* (Vol. 29). London/Berkeley/Los Angeles: University of California Press. Collected translation from different original work, from 1909 to 1920.

Wilensky, U., & Rand, W. (2007). Making models match: Replicating an agent-based model. *Journal of Artificial Societies and Social Simulation, 10*(4), 2.

Wimsatt, W. (2000). Emergence as non-aggregativity and the biases of reductionism. *Foundations of Science, 5*(3), 269–297.

Wimsatt, W. C. (1986). *Forms of aggregativity* (pp. 259–291). Dordrecht: Reidel.

Wimsatt, W. C. (2006a). Aggregate, composed, and evolved systems: Reductionistic heuristics as means to more holistic theories. *Biology & Philosophy, 21*(5), 667–702.

Wimsatt, W. C. (2006b). Reductionism and its heuristics: Making methodological reductionism honest. *Synthese, 151*(3), 445–475.

Wolfram, S. (1983). Statistical mechanics of cellular automata. *Reviews of Modern Physics, 55*(3), 601–644.

Wolfram, S. (2002). *A new kind of science.* Champaign, IL: Wolfram Media Inc.

Wright, S. (1932). The roles of mutation, inbreeding, crossbreeding, and selection in evolution. In *Proceedings of the Sixth International Congress on Genetics.*

Chapter 3
Agent-Based Models as Markov Chains

This chapter spells out the most important theoretical ideas developed in this book. However, it begins with an illustrative introductory description of agent-based models (ABMs) in order to provide an intuition for what follows. It then shows for a class of ABMs that, at the micro level, they give rise to random walks on regular graphs (Sect. 3.2). The transition from the micro to the macro level is formulated in Sect. 3.3. When a model is observed in terms of a certain system property, this effectively partitions the state space of the micro chains such that micro configurations with the same observable value are projected into the same macro state. The conditions for the projected process to be again a Markov chain are given which relates the symmetry structure of the micro chains to the partition induced by macroscopic observables. We close with a simple example that will be discussed further in the next chapter.

3.1 Basic Ingredients of Agent-Based Models

Roughly speaking, an ABM is a set of autonomous agents which interact according to relatively simple interactions rules with other agents and the environment. The agents themselves are characterized (or modeled) by a set of attributes some of which may change over time. Interaction rules specify the agent behavior with respect to other agents in the social environment and in some models there are also rules for the interaction with an external environment. Accordingly, the environment in an AB simulation is sometimes a model of a real physical space in which the agents move and interact upon encounter, in other models interaction relations between the agents are defined by an agent interaction network and the resulting neighborhood structure.

© Springer International Publishing Switzerland 2016
S. Banisch, *Markov Chain Aggregation for Agent-Based Models*,
Understanding Complex Systems, DOI 10.1007/978-3-319-24877-6_3

Fig. 3.1 Caricature of an agent

In the simulation of an ABM the interaction process is iterated and the repeated application of the rules gives rise to the time evolution. There are different ways in which this update may be conceived and implemented. As virtually all ABMs are made to be simulated on a computer, I think it is reasonable to add to the classic threefold characterization of AB systems as "agents plus interactions plus environment" a time-component because different modes of event scheduling can be of considerable importance.

3.1.1 Agents as Elementary Units

In this work, we deal with agents that are characterized by a finite set of attributes. The agent in the example shown in Fig. 3.1, for instance, can be described by a four-dimensional vector encoding the four different attributes from top to the bottom. In the sequel we will denote the state of an agent i as x_i. Let us assume that, in this example, for each of the four features there are two alternatives: blank or covered. Then we could encode its state from the top to the bottom as $x_i = (\blacksquare\square\blacksquare\square)$, $\blacksquare$ accounting for "covered" and $\square$ for "blank". It is clear that, in this case, there are $2^4 = 16$ possible agent states and we shall refer to this set as *attribute space* and denote it by $\mathbf{S} = \{\blacksquare, \square\}^4$.

For the purposes of this work, the meaning of the content of such attributes is not important because the interpretation depends on the application for which the agent model is designed. It could account for the behavioral strategies with regard to four different dimensions of an agent's live, it could be words or utterances that the agent prefers in a communication with others, or represent a genetic disposition. Consequently, x_i may encode static agent attributes or qualities that change in the life-time of the agent, or a mixture of static and dynamic features.

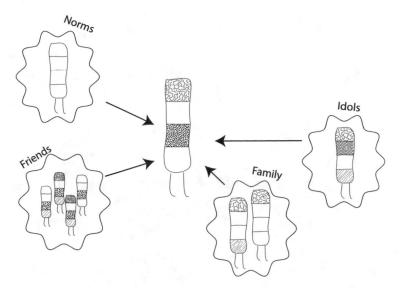

Fig. 3.2 A social agent and its environment

AB simulation is usually an attempt to analyze the behavior of an entire population of agents as it follows from many individual decisions. Therefore, there is actually a number of N agents each one characterized by a state $x_j \in S$. We shall denote the configuration of N agents by $\mathbf{x} = (x_1, \ldots, x_N)$ and call this an *agent profile* or *agent configuration*.

3.1.2   The Environment

For the moment, we keep our eye on a single agent and consider environmental aspects an agent may take into account for its decisions (Fig. 3.2). As noted earlier, the environment can be a model of real physical space in which the agent moves around according to some movement rules and where interaction with other individuals occurs whenever these agents encounter in the physical space. But environment is actually a more abstract concept in AB modeling. It also accounts for the agent's social environment, its friends and family, as well as for social norms, idols or fads brought about by television. In a biological context the environment might be modeled by a fitness function which assigns different reproduction chances to different agent attributes x_i.

One of the most important aspects in AB modeling is the introduction of social relations between the agents. Family structures and friendship relations are usually included by means of a graph $G = (\mathbf{N}, \mathbf{E})$, the so-called social network. Here $\mathbf{N}$ denotes the set of agents and $\mathbf{E}$ is the set of connections (i, j) between the agents. These connections, called edges, can be weighted to account for the strength of the relation between agent i and j and negative values might even be taken to

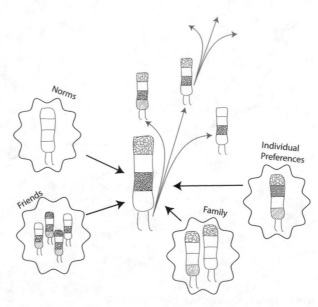

Fig. 3.3 Interaction and iteration involve indeterminism and stochasticity. Therefore, there are several possible future states to which an agent may evolve in one step

model adverse relations. Very often, the probability that two agents are part of the same interaction event depends directly on their connectivity in G. In fact, many models, especially simple physics-inspired models of social dynamics, take into account only a social interaction network and leave other environmental aspects out of consideration.

3.1.3   Interaction Rules

In an interaction event, typically, an agent has to take a decision on the basis of the information within its environment. This includes a set of other agents, friends, family, with which the agent is connected as well as global information about norms, and possibly, internalized individual preferences. Each decision corresponds to an update of the agent's state $x_i \rightarrow y_i$ where we use x_i to denote the agent state before the interaction takes place and y_i to denote the updated state (Fig. 3.3).

Usually, an agent in a specific situation has several well-defined behavioral options. Although in some sophisticated models agents are endowed with the capacity of evaluating the efficiency of these options, it is an important mark of ABMs that this evaluation is based on incomplete information and not perfect, and therefore the choice an agent takes involves a level of uncertainty. That is, a probability is assigned to the different options and the choice is based on those probabilities. This means that an agent in state x_i may end up after the interaction

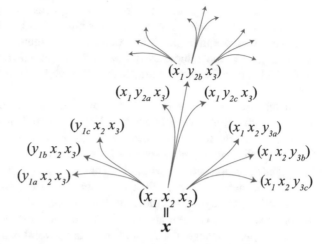

Fig. 3.4 Possible paths in a small system of three agents (labeled by 1, 2, 3) where every agent has three alternative options (labeled by a, b, c)

$$(x_1 y_{2b} x_3)$$
$$(x_1 y_{2a} x_3) \qquad (x_1 y_{2c} x_3)$$
$$(y_{1c} x_2 x_3) \qquad (x_1 x_2 y_{3a})$$
$$(y_{1b} x_2 x_3) \qquad (x_1 x_2 y_{3b})$$
$$(y_{1a} x_2 x_3) \qquad (x_1 x_2 y_{3c})$$
$$(x_1 x_2 x_3)$$
$$\parallel$$
$$x$$

in different states $y_i, y_i', y_i'', \ldots$ The indeterminism introduced in this way is an essential difference to neoclassical game-theoretic models and rational choice theory. And it is the reason why Markov chain theory is such a good candidate for the mathematical formalization of the AB dynamics.

3.1.4   Iteration Process

The conceptual design of an ABM is mainly concerned with a proper definition of agents, their interaction rules and the environment in which they are situated. In order to study the time evolution of such a system of interdependent agents, however, it is also necessary to define how the system proceeds from one time step to the other. As virtually all ABMs are simulation models implemented on a computer, it is an inherent part of the modeling task to specify the order in which events take place during an update of the system.

A typical procedure is to first choose an agent at random (say agent i). The current agent state x_i along with all the information this agent has about his environment defines the actual situation of the agent and determines the different behavioral options. If, in this situation, there is more than one option available to the agent, in a second step, one of these options has to be chosen with a certain probability. In this light, the update of an AB system can be seen as a stochastic choice out of a set of deterministic options, where stochastic elements are involved first into the agent choice and second into the selection of one out of several well-defined alternatives.

This procedure is illustrated for a small system of three agents in Fig. 3.4. The current agent profile is $x = (x_1 x_2 x_3)$. To proceed to the next time step, first, one of the agents is chosen to update its state with some probability. So the new configuration of the system (denoted as y) might differ from x in the first ($x_1 \rightarrow y_1$), the second ($x_2 \rightarrow y_2$), or the third ($x_3 \rightarrow y_3$) position. As every agent himself has three different behavioral alternatives chosen with a certain probability (as in Fig. 3.3), there are three paths for each potential agent ($x_1 \rightarrow y_{1a}$ or $x_1 \rightarrow y_{1b}$ or $x_1 \rightarrow y_{1c}$). As a whole, there are thus 9 ($= 3 \times 3$) possible future agent configurations **y** to which the update process may lead with a well-defined probability after a single step.

In the update scheme described above the agents are updated one after the other and therefore this scheme is called sequential or sometime asynchronous update. A single time step corresponds in this scheme to a single interaction event. An alternative update scheme is synchronous or simultaneous update where the agents are updated "in parallel". That is, given a system profile **x**, all agents are chosen, determine and select their behavioral options at the same time. The transition structure becomes more complex in that case mainly because the number of possible future configurations **y** is large compared to the asynchronous case since all agents change at once and there are several paths for each agent. In our example system of three agents each with three different options, the number of possible future states **y** is 27 ($= 3^3$). Most ABMs, however, have been implemented using the sequential update scheme, may be because the sequential philosophy of traditional programming languages made it more convenient. In this work, we will also concentrate on the sequential scheme.

3.2 The Micro Level

3.2.1 The Grammar of an Agent-Based Model

Let us consider an abstract ABM with finite configuration space $\Sigma = \mathbf{S}^N$ (meaning that there are N agents with attributes $x_i \in \mathbf{S}$). Any iteration of the model (any run of the ABM algorithm) maps a configuration $\mathbf{x} \in \Sigma$ to another configuration $\mathbf{y} \in \Sigma$. In general, the case that no agent changes such that $\mathbf{x} = \mathbf{y}$ is also possible. Let us denote such a mapping by $F_z : \Sigma \rightarrow \Sigma$ and denote the set of all possible mappings by $\mathscr{F}$. Notice that any element of $\mathscr{F}$ can be seen as a word of length $|\Sigma|$ over an $|\Sigma|$-ary alphabet, and there are $|\Sigma|^{|\Sigma|}$ such words (Flajolet and Odlyzko 1990, p. 3).

Any $F_z \in \mathscr{F}$ induces a directed graph (Σ, F_z) the nodes of which are the elements in Σ (i.e., the agent configurations) and edges the set of ordered pairs $(\mathbf{x}, F_z(\mathbf{x}))$, $\forall \mathbf{x} \in \Sigma$. Such a graph is called functional graph of F_z because it displays the functional relations of the map F_z on Σ. That is, it represents the logical paths induced by F_z on the space of configurations for any initial configuration $\mathbf{x}$.

Each iteration of an ABM can be thought of as a stochastic choice out of a set of deterministic options. For an ABM in a certain configuration $\mathbf{x}$, there are usually several options (several $\mathbf{y}$) to which the algorithm may lead with a well-defined probability (see Sect. 3.1). Therefore, in an ABM, the transitions between the different configurations $\mathbf{x}, \mathbf{y}, \ldots \in \Sigma$ are not defined by one single map F,

Fig. 3.5 Possible paths from
configuration $\mathbf{x} = (\square \blacksquare \blacksquare)$
in a small VM of three agents

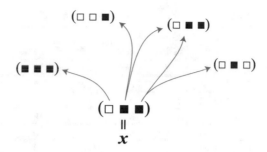

but there is rather a subset $\mathscr{F}_Z \subset \mathscr{F}$ of maps out of which one map is chosen at
each time step with certain probability. Let us assume we know all the mappings
$\mathscr{F}_Z = \{F_1, \ldots, F_z, \ldots, F_n\}$ that are realized by the ABM of our interest. With this,
we are able to define a functional graph representation by $(\Sigma, \mathscr{F}_Z)$ which takes as
the nodes all elements of Σ (all agent configurations) and an arc $(\mathbf{x}, \mathbf{y})$ exists if there
is at least one $F_z \in \mathscr{F}_Z$ such that $F_z(\mathbf{x}) = \mathbf{y}$. This graph defines the "grammar" of
the system for it displays all the logically possible transitions between any pair of
configurations of the model.

Consider the VM with three agents as an example. In the VM agents have two
possible states ($\mathbf{S} = \{\sqcap, \blacksquare\}$) and the configuration space for a model of three agents
is $\Sigma = \{\square, \blacksquare\}^3$. In the iteration process, one agent i is chosen at random along
with one of its neighbors j and agent i imitates the state of j. This means that $y_i = x_j$
after the interaction event. Notice that once an agent pair (i, j) is chosen the update
is defined by a deterministic map $\mathbf{u} : \mathbf{S}^2 \to \mathbf{S}$. Stochasticity enters first because
of the random choice of i and second through the random choice of one agent in
the neighborhood. Let us look at an example with three agents in the configuration
$\mathbf{x} = (\square \blacksquare \blacksquare)$. If the first agent is chosen ($i = 1$ and $x_1 = \square$) then this agent will
certainly change state to $y_1 = \blacksquare$ because it will in any case meet a black agent. For
the second and the third agent ($i = 2$ or $i = 3$) the update result depends on whether
one or the other neighbor is chosen because they are in different states. Noteworthy,
different agent choices may lead to the same configuration. Here, this is the case if
the agent pair $(2, 3)$ or $(3, 2)$ is chosen in which case the agent (2 or 3) does not
change its state because $x_2 = x_3$. Therefore we have $\mathbf{y} = \mathbf{x}$ and there are two paths
realizing that transition (Fig. 3.5).

In practice, the explicit construction of the entire functional graph may rapidly
become a tedious task due to the huge dimension of the configuration space and
the fact that one needs to check if $F_z(\mathbf{x}) = \mathbf{y}$ for each mapping $F_z \in \mathscr{F}_Z$ and all
pairs of configurations $\mathbf{x}, \mathbf{y}$. On the other hand, the main interest here is a theoretical
one, because, as a matter of fact, a representation as a functional graph of the form
$\Gamma = (\Sigma, \mathscr{F}_Z)$ exists for any model that comes in form of a computer algorithm.
It is therefore a quite general way of formalizing ABMs and, as we will see in the
sequel, allows under some conditions to verify the Markovianity of the models at
the micro level.

Table 3.1 $\mathscr{F}_Z$ for the VM with three agents

		a ■■■	b ■■□	c ■□■	d □■■	e ■□□	f □■□	g □□■	h □□□
z	(i,j)								
1	$(1,2)$	a	b	g	a	h	b	g	h
2	$(1,3)$	a	f	c	a	h	f	c	h
3	$(2,1)$	a	b	a	g	b	h	g	h
4	$(3,1)$	a	a	c	f	c	f	h	h
5	$(2,3)$	a	e	a	d	e	h	d	h
6	$(3,2)$	a	a	e	d	e	d	h	h

If we really want to construct the "grammar" of an ABM explicitly this requires the dissection of stochastic and deterministic elements of the iteration procedure of the model. As an example, let us consider again the VM for which such a dissection is not difficult. In the VM, the random part consists of the choice of two connected agents (i,j). Once this choice is made we know that $y_i = x_j$ by the interaction rule. This is sufficient to derive the functional representation of the VM, because we need only to check one by one for all possible choices (i,j) which transitions this choice induces on the configuration space. For a system of three agents, with all agents connected to the other two, the set of functions $\mathscr{F}_Z = \{F_1, \ldots, F_z, \ldots, F_n\}$ is specified in Table 3.1. Notice that with three agents, there are 8 possible configurations indexed here by $a, b, \ldots, h$. Moreover, there are 6 possible choices for (i,j) such that $\mathscr{F}_Z$ consists of $n = 6$ mappings.

Each row of the table represents a mapping $F_z : \Sigma \to \Sigma$ by listing to which configuration **y** the respective map takes the configurations a to h. The first row, to make an example, represents the choice of the agent pair $(1,2)$. The changes this choice induces depend on the actual agent configuration **x**. Namely, for any **x** with $x_1 = x_2$ we have $F_1(\mathbf{x}) = F_{(1,2)}(\mathbf{x}) = \mathbf{x}$. So the configurations a, b, g, h are not changed by $F_{(1,2)}$. For the other configurations it is easy to see that (■□■) $\to$ (□□■) $(c \to g)$, (□■■) $\to$ (■■■) $(d \to a)$, (■□□) $\to$ (□□□) $(e \to h)$, and (□■□) $\to$ (■■□) $(f \to b)$. Notice that the two configurations (□□□) and (■■■) with all agents equal are not changed by any map and correspond therefore to the final configurations of the VM.

In Fig. 3.6, the complete functional graph $\Gamma = (\Sigma, \mathscr{F}_Z)$ of the VM with three agents is shown. This already gives us some important information about the behavior of the VM such as the existence of two final configurations with all agents in the same state. We also observe that the VM iteration gives rise to a very regular functional graph, namely, the N-dimensional hypercube. In what follows, we show how to derive the respective transition probabilities associated to the arrows in Fig. 3.6.

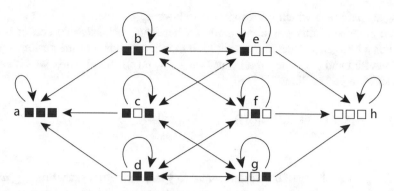

Fig. 3.6 Grammar of the VM with three agents

3.2.2 From Functional Graphs to Markov Chains

A functional graph $\Gamma = (\Sigma, \mathscr{F}_Z)$ defines the "grammar" of an ABM in the sense that it shows all possible transitions enabled by the model. It is the first essential step in the construction of the Markov chain associated with the ABM at the micro level because there is a non-zero transition probability only if there is an arrow in the functional graph. Consequently, all that is missing for a Markov chain description is the computation of the respective transition probabilities.

For a class of models, including the VM, this is relatively simple because we can derive a random mapping representation (Levin et al. 2009, pp. 6/7) directly from the ABM rules. Namely, if $F_{z_1}, F_{z_2}, \ldots$ is a sequence of independent random maps, each having the same distribution ω, and $S_0 \in \Sigma$ has distribution μ_0, then the sequence $S_0, S_1, \ldots$ defined by

$$S_t = F_{z_t}(S_{t-1}), t \geq 1 \tag{3.1}$$

is a Markov chain on Σ with transition matrix $\hat{P}$:

$$\hat{P}(\mathbf{x}, \mathbf{y}) = \mathbf{Pr}_\omega[z, F_z(\mathbf{x}) = \mathbf{y}]; \mathbf{x}, \mathbf{y} \in \Sigma. \tag{3.2}$$

Conversely (Levin et al. 2009), any Markov chain has a random map representation (RMR). Therefore, in that case, (3.1) and (3.2) may be taken as an equivalent definition of a Markov chain. This is particularly useful in our case, because it shows that an AB simulation models which can be described as above is, from a mathematical point of view, a Markov chain. This includes several models described in Izquierdo et al. (2009).

In the VM, the separation of stochastic and deterministic elements is clear-cut and therefore a random mapping representation is obtained easily. As already shown in Table 3.1, we can use the possible agent choices (i, j) directly to index the collection of maps $F_{(i,j)} \in \mathscr{F}_Z$. We denote as $\omega(i, j)$ the probability of choosing the agent pair (i, j) which corresponds to choosing the map $F_{(i,j)}$. It is clear that we can proceed in this way in all models where the stochastic part concerns only the choice of agents. Then, the distribution ω is independent of the current system configuration and the same for all times $(\omega(z_t) = \omega(z))$. In this case, we obtain for the transition probabilities

$$\hat{P}(\mathbf{x}, \mathbf{y}) = \mathbf{Pr}_\omega[(i,j), F_{(i,j)}(\mathbf{x}) = \mathbf{y}] = \sum_{\substack{(i,j): \\ F_{(i,j)}(\mathbf{x}) = \mathbf{y}}} \omega(i,j). \qquad (3.3)$$

That is, the probability of transition from $\mathbf{x}$ to $\mathbf{y}$ is the conjoint probability $\sum \omega(i,j)$ of choosing an agent pair (i,j) such that the corresponding map takes $\mathbf{x}$ to $\mathbf{y}$ (i.e., $F_{(i,j)}(\mathbf{x}) = \mathbf{y}$).

3.2.3 Single-Step Dynamics and Random Walks on Regular Graphs

In this thesis, we focus on a class of models which we refer to as *single-step dynamics*. They are characterized by the fact that only one agent changes at a time step.[1] Notice that this is very often the case in ABMs with a sequential update scheme and that sequential update is, as a matter of fact, the most typical iteration scheme in ABMs. In terms of the "grammar" of these models, this means that non-zero transition probabilities are only possible between system configuration that differ in at most one position. And this gives rise to random walks on regular graphs.

Consider a set of N agents each one characterized by individual attributes x_i that are taken in a finite list of possibilities $\mathbf{S} = \{1, \ldots, \delta\}$. In this case, the space of possible agent configurations is $\Sigma = \mathbf{S}^N$. Consider further a deterministic update function $\mathbf{u} : \mathbf{S}^r \times \Lambda \to \mathbf{S}$ which takes configuration $\mathbf{x} \in \Sigma$ at time t to configuration $\mathbf{y} \in \Sigma$ at $t+1$ by

$$y_i = \mathbf{u}(x_i, x_j, \ldots, x_k, \lambda). \qquad (3.4)$$

To go from one time step to the other in agent systems, usually, an agent i is chosen first to perform a step. The decision of i then depends on its current state (x_i) and the attributes of its neighbors $(x_j, \ldots, x_k)$. The finite set Λ accounts for a possible stochastic part in the update mechanism such that different behavioral options are implemented by different update functions $\mathbf{u}(\ldots, \lambda_1)$, $\mathbf{u}(\ldots, \lambda_2)$ etc. Notice that for the case in which the attributes of the agents $(x_i, x_j, \ldots, x_k)$ uniquely determine the agent decision we have $\mathbf{u} : \mathbf{S}^r \to \mathbf{S}$ which strongly resembles the update rules implemented in cellular automata (CA).

---

[1]Notice that a slightly more general class of models has been considered in Banisch et al. (2012).

As opposed to classical CA, however, a sequential update scheme is used in the class of models considered here. In the iteration process, first, a random choice of agents along with a λ to index the possible behavioral options is performed with probability $\omega(i, j, \ldots, k, \lambda)$. This is followed by the application of the update function which leads to the new state of agent i by Eq. (3.4).

Due to the sequential application of an update rule of the form $\mathbf{u} : \mathbf{S}^r \times \Lambda \to \mathbf{S}$ only one agent (namely agent i) changes at a time so that all elements in $\mathbf{x}$ and $\mathbf{y}$ are equal except that element which corresponds to the agent that was updated during the step from $\mathbf{x}$ to $\mathbf{y}$. Therefore, $x_j = y_j$, $\forall j \neq i$ and $x_i \neq y_i$. We call $\mathbf{x}$ and $\mathbf{y}$ adjacent and denote this by $\mathbf{x} \overset{i}{\sim} \mathbf{y}$.

It is then also clear that a transition from $\mathbf{x}$ to $\mathbf{y}$ is possible if $\mathbf{x} \sim \mathbf{y}$. Therefore, the adjacency relation $\sim$ defines the "grammar" Γ_{SSD} of the entire class of single-step models. Namely, the existence of a map F_z that takes $\mathbf{x}$ to $\mathbf{y}$, $\mathbf{y} = F_z(\mathbf{x})$, implies that $\mathbf{x} \overset{i}{\sim} \mathbf{y}$ for some $i \in \{1, \ldots, N\}$. This means that any ABM that belongs to the class of single-step models performs a walk on Γ_{SSD} or on a subgraph of it.

Let us briefly consider the structure of the graph Γ_{SSD} associated to the entire class of single-step models. From $\mathbf{x} \overset{i}{\sim} \mathbf{y}$ for $i = 1, \ldots, N$ we know that for any $\mathbf{x}$, there are $(\delta - 1)N$ different vectors $\mathbf{y}$ which differ from $\mathbf{x}$ in a single position, where δ is the number of possible agent attributes. Therefore, Γ_{SSD} is a regular graph with degree $(\delta - 1)N + 1$, because in our case, the system may loop by $y_i = x_i$. As a matter of fact, our definition of adjacency as "different in one position of the configuration" is precisely the definition of so-called Hamming graphs which tells us that $\Gamma_{SSD} = H(N, \delta)$ (with loops). In the case of the VM, where $\delta = 2$ we find $H(N, 2)$ which corresponds to the N-dimensional hypercube.

As before, the transition probability matrix of the micro chain is denoted by $\hat{P}$ with $\hat{P}(\mathbf{x}, \mathbf{y})$ being the probability for the transition from $\mathbf{x}$ to $\mathbf{y}$. The previous considerations tell us that non-zero transition probabilities can exist only between two configurations that are linked in $H(N, d)$ plus the loop ($\hat{P}(\mathbf{x}, \mathbf{x})$). Therefore, each row of $\hat{P}$ contains no more than $\delta N + 1$ non-zero entries. In the computation of $\hat{P}$ we concentrate on pairs of adjacent configurations. For $\mathbf{x} \overset{i}{\sim} \mathbf{y}$ with $x_i \neq y_i$ we have

$$\hat{P}(\mathbf{x}, \mathbf{y}) = \sum_{\substack{(i,j,\ldots,k,\lambda): \\ y_i = \mathbf{u}(x_i, x_j, \ldots, x_k, \lambda)}} \omega(i, j, \ldots, k, \lambda) \tag{3.5}$$

which is the conjoint probability to choose agents and a rule $(i, j, \ldots, k, \lambda)$ such that the ith agent changes its attribute by $y_i = \mathbf{u}(x_i, x_j, \ldots, x_k, \lambda)$. For the probability that the model remains in $\mathbf{x}$, $\hat{P}(\mathbf{x}, \mathbf{x})$, we have

$$\hat{P}(\mathbf{x}, \mathbf{x}) = 1 - \sum_{\mathbf{y} \sim \mathbf{x}} \hat{P}(\mathbf{x}, \mathbf{y}). \tag{3.6}$$

Table 3.2 Update rules $y_i = \mathbf{u}(x_i, x_j)$ for the voter model (VM), anti-ferromagnetic coupling (AC) and diffusion (DF)

	x_j	x_j	VM	■	□	AC	■	□	DF	■	□
x_i	y_i	y_i	■	■	□	■	■	□	■	■	■
x_i	y_i	y_i	□	■	□	□	□	□	■	■	□

Equation (3.5) makes visible that the probability distribution ω plays the crucial role in the computation of the elements of $\hat{P}$.

The VM is a very simple instance of single-step dynamics. The update function is given by $y_i = \mathbf{u}(x_i, x_j) = x_j$ and the stochastic part of the model concerns only the choice of an agent pair (i, j) with probability $\omega(i, j)$. For adjacent configuration with $\mathbf{x} \overset{i}{\sim} \mathbf{y}$, Eq. (3.5) simplifies to

$$\hat{P}(\mathbf{x}, \mathbf{y}) = \sum_{j:(y_i = \mathbf{u}(x_i, x_j))} \omega(i, j) = \sum_{j:(y_i = x_j)} \omega(i, j) \tag{3.7}$$

Notice that (3.7) is applicable to all ABMs in which first an agent pair (i, j) is chosen at random and second a deterministic update rule $y_i = \mathbf{u}(x_i, x_j)$ defines the outcome of the interaction between i and j. For a binary attribute space $\mathbf{S} = \{\square, \blacksquare\}$ some possible update rules $\mathbf{u} : \mathbf{S} \times \mathbf{S} \rightarrow \mathbf{S}$ are shown in Table 3.2 below.

3.3 Macrodynamics, Projected Systems and Observables

3.3.1 Micro and Macro in Agent-Based Models

What do we look at when we analyze an ABM? Typically, we try to capture the dynamical behavior of a model by studying the time evolution of parameters or indicators that inform us about the global state of the system. Although, in some cases, we might understand the most important dynamical features of a model by looking at repeated visualizations of all details of the agent system through time, basic requirements of the scientific method will eventually enforce a more systematic analysis of the model behavior in the form of systematic computational experiments and "extensive sensitivity analysis" (Epstein 2006, p. 28). In this, there is no other choice than to leave the micro level of all details and to project the system behavior or state onto global structural indicators representing the system as a whole. In many cases, a description like that will even be desired, because the focus of attention in ABMs, the facts to be explained, are usually at a higher macroscopic level beyond the microscopic description. In fact, the search for microscopic foundations for macroscopic regularities has been an integral motivation for the development of AB research (see Macy and Willer 2002; Squazzoni 2008).

It is characteristic of any such macroscopic system property that it is invariant with respect to certain details of the agent configuration. In other words, any observation defines, in effect, a many-to-one relation by which sets of micro configurations with the same observable value are subsumed into the same macro state. Consider the population dynamics in the sugarscape model by Epstein and Axtell (1996) as an example. The macroscopic indicator is, in this case, the number of agents N. This aggregate value is not sensitive with respect to the exact positions (the sites) at which the agents are placed, but only to how many sites are occupied. Consequently, there are many possible configurations of agent occupations in the sugarspace with an equal number of agents N and all of them correspond to the same macro state. Another slightly more complicated example is the skewed wealth distribution in the sugarscape model. It is not important which agents contribute to each specific wealth (sugar) level, but only how many there are in each level. This describes how macro descriptions of ABMs are related to observations, system properties, order parameters and structural indicators, and it also brings into the discussion to the concepts of aggregation and decomposition.

Namely, aggregation is one way (in fact, a very common one) of realizing such a many-to-one mapping from micro-configurations to macroscopic system properties and observables. For simple models of opinion dynamics inspired by spin physics, for instance, it is very common to use the average opinion—due to the spin analogy often called "system magnetization"—as an order parameter and to study the system behavior in this way. Magnetization, computed by summation over the spins and division by the total number of spins, is a true aggregative measure. Magnetization levels or values are then used to classify spin or opinion configurations, such that those configurations with the same magnetization value correspond to the same macro state. This many-to-one mapping of sets of micro configurations onto macro states automatically introduces a decomposition of the state space at the micro level Σ.

3.3.2 Observables, Partitions and Projected Systems

The formulation of an ABM as a Markov chain developed in the previous section allows a formalization of this micro-macro link in terms of projections. Namely, a projection of a Markov chain with state space Σ is defined by a new state space $\mathbf{X}$ and a projection map Π from Σ to $\mathbf{X}$. The meaning of the projection Π is to lump sets of micro configurations in Σ according to some macro property in such a way that, for each $X \in \mathbf{X}$, all the configurations of Σ in $\Pi^{-1}(\mathbf{X})$ share the same property.

Therefore, such projections are important when catching the macroscopic properties of the corresponding ABM because they are in complete correspondence with a classification based on an observable property of the system. To see how this correspondence works let us suppose that we are interested in some factual property of our agent-based system. This means that we are able to assign to each configuration the specific value of its corresponding property. Regardless of the kind of value used to specify the property (qualitative or quantitative), the set $\mathbf{X}$ needed to describe the configurations with respect to the given property is a finite set, because the set of all configurations is also finite. Let then $\phi : \Sigma \to \mathbf{X}$ be the function that assigns to any configuration $\mathbf{x} \in \Sigma$ the corresponding value of the considered property. It is natural to call such ϕ an observable of the system. Now, any observable of the system naturally defines a projection Π by lumping the set of all the configurations with the same ϕ value. Conversely any (projection) map Π from Σ to $\mathbf{X}$ defines an observable ϕ with values in the image set $\mathbf{X}$. Therefore these two ways of describing the construction of a macro-dynamics are equivalent and the choice of one or the other point of view is just a matter of taste.

The price to pay in passing from the micro to the macro-dynamics in this sense (Kemeny and Snell 1976; Chazottes and Ugalde 2003) is that the projected system is, in general, no longer a Markov chain: long memory (even infinite) may appear in the projected system. This "complexification" of the macro dynamics with respect to the micro dynamics is a fingerprint of dynamical emergence in agent-based and other computational models (cf. Humphreys 2008).

3.3.3 Lumpability and Symmetry

Under certain conditions, the projection of a Markov chain $(\Sigma, \hat{P})$ onto a coarse-grained partition $\mathbf{X}$, obtained by aggregation of states, is still a Markov chain. In Markov chain theory this is known as lumpability (or strong lumpability), and necessary and sufficient conditions for this to happen are known. Let us restate the respective Theorem 6.3.2 of Kemeny and Snell (1976) using our notations, where Σ denotes the configuration space of the micro chain and $\hat{P}$ the respective transition matrix, and $\mathbf{X} = (X_1, \ldots, X_r)$ is a partition of Σ. Let $\hat{p}_{\mathbf{x}Y} = \sum_{\mathbf{y} \in Y} \hat{P}(\mathbf{x}, \mathbf{y})$ denote the conjoint probability for $\mathbf{x} \in \Sigma$ to go to the set of elements $\mathbf{y} \in Y$ where $Y \subseteq \Sigma$ is a subset of the configuration space.

Theorem 3.1 (Kemeny and Snell 1976, p. 124) *A necessary and sufficient condition for a Markov chain to be lumpable with respect to a partition $\mathbf{X} = (X_1, \ldots, X_r)$ is that for every pair of sets X_i and X_j, $\hat{p}_{\mathbf{x}X_j}$ have the same value for every $\mathbf{x}$ in X_i. These common values $\{\hat{p}_{ij}\}$ form the transition matrix for the lumped chain.*

In general it may happen that, for a given Markov chain, some projections are Markov and others not. Therefore a judicious choice of the macro properties to be studied may help the analysis.

In order to establish the lumpability in the cases of interest we shall use symmetries of the model. For further convenience, we state a result for which the proof is easily given Theorem 6.3.2 of Kemeny and Snell (1976):

Theorem 3.2 *Let $(\Sigma, \hat{P})$ be a Markov chain and $\mathbf{X} = (X_1, \ldots, X_n)$ a partition of Σ. Suppose that there exists a group $\mathscr{G}$ of bijections on Σ that preserve the partition*

($\forall \mathbf{x} \in X_i$ and $\forall \hat{\sigma} \in \mathcal{G}$ we have $\hat{\sigma}(\mathbf{x}) \in X_i$). If the Markov transition probability $\hat{P}$ is symmetric with respect to $\mathcal{G}$,

$$\hat{P}(\mathbf{x}, \mathbf{y}) = \hat{P}(\hat{\sigma}(\mathbf{x}), \hat{\sigma}(\mathbf{y})) : \forall \hat{\sigma} \in \mathcal{G}, \tag{3.8}$$

the partition $(X_1, \ldots, X_n)$ is (strongly) lumpable.

Proof For the proof it is sufficient to show that any two configurations $\mathbf{x}$ and $\mathbf{x}'$ with $\mathbf{x}' = \hat{\sigma}(\mathbf{x})$ satisfy

$$\hat{p}_{\mathbf{x}Y} = \sum_{\mathbf{y} \in Y} \hat{P}(\mathbf{x}, \mathbf{y}) = \sum_{\mathbf{y} \in Y} \hat{P}(\mathbf{x}', \mathbf{y}) = \hat{p}_{\mathbf{x}'Y} \tag{3.9}$$

for all $Y \in \mathbf{X}$. Consider any two subsets $X, Y \in \mathbf{X}$ and take $\mathbf{x} \in X$. Because $\mathcal{G}$ preserves the partition it is true that $\mathbf{x}' \in X$. Now we have to show that Eq. (3.9) holds. First the probability for $\mathbf{x}' = \hat{\sigma}(\mathbf{x})$ to go to an element $\mathbf{y} \in Y$ is

$$\hat{p}_{\hat{\sigma}(\mathbf{x})Y} = \sum_{\mathbf{y} \in Y} \hat{P}(\hat{\sigma}(\mathbf{x}), \mathbf{y}). \tag{3.10}$$

Because the $\hat{\sigma}$ are bijections that preserve the partition $\mathbf{X}$ we have $\hat{\sigma}(Y) = Y$ and there is for every $\mathbf{y} \in Y$ exactly one $\hat{\sigma}(\mathbf{y}) \in Y$. Therefore we can substitute

$$\hat{p}_{\hat{\sigma}(\mathbf{x})Y} = \sum_{\mathbf{y} \in Y} \hat{P}(\hat{\sigma}(\mathbf{x}), \hat{\sigma}(\mathbf{y})) = \sum_{\mathbf{y} \in Y} \hat{P}(\mathbf{x}, \mathbf{y}) = \hat{p}_{\mathbf{x}Y}, \tag{3.11}$$

where the second equality comes by the symmetry condition (3.8) that $\hat{P}(\mathbf{x}, \mathbf{y}) = \hat{P}(\hat{\sigma}(\mathbf{x}), \hat{\sigma}(\mathbf{y}))$.

The usefulness of the conditions for lumpability stated in Theorem 3.2 becomes apparent recalling that AB simulations can be seen as random walks on regular graphs defined by the functional graph or "grammar" of the model $\Gamma = (\Sigma, \mathscr{F}_Z)$. The full specification of the random walk $(\Sigma, \hat{P})$ is obtained by assigning transition probabilities to the connections in Γ and we can interpret this as a weighted graph. The regularities of $(\Sigma, \hat{P})$ are captured by a number of non-trivial automorphisms which, in the case of ABMs, reflect the symmetries of the models.

In fact, Theorem 3.2 allows to systematically exploit the symmetries of an agent model in the construction of partitions with respect to which the micro chain is lumpable. Namely, the symmetry requirement in Theorem 3.2, that is, Eq. (3.8), corresponds precisely to the usual definition of automorphisms of $(\Sigma, \hat{P})$. The set of all permutations $\hat{\sigma}$ that satisfy (3.8) corresponds then to the automorphism group of $(\Sigma, \hat{P})$.

Lemma 3.1 *Let $\mathcal{G}$ be the automorphism group of the micro chain $(\Sigma, \hat{P})$. The orbits of $\mathcal{G}$ define a lumpable partition $\mathbf{X}$ such that every pair of micro configurations $\mathbf{x}, \mathbf{x}' \in \Sigma$ for which $\exists \hat{\sigma} \in \mathcal{G}$ such that $\mathbf{x}' = \hat{\sigma}(\mathbf{x})$ belong to the same subset $X_i \in \mathbf{X}$.*

Note 3.1 Lemma 3.1 actually applies to any $\mathcal{G}$ that is a proper subgroup of the automorphism group of $(\Sigma, \hat{P})$. The basic requirement for such a subset $\mathcal{G}$ to be a group is that be closed under the group operation which establishes that $\hat{\sigma}(X_i) = X_i$. With the closure property, it is easy that any such subgroup $\mathcal{G}$ defines a lumpable partition in the sense of Theorem 3.2.

3.4 From Automorphisms to Macro Chains

In this section we illustrate the previous ideas at the example of three state single-step dynamics. Consider a system of N agents each one characterized by an attribute $x_i \in \{a, b, c\}$, that is $\delta = 3$. As discussed in Sect. 3.2.3, the corresponding graph Γ encoding all the possible transitions is the Hamming graph $H(N, 3)$. The nodes $\mathbf{x}, \mathbf{y}$ in $H(N, 3)$ correspond to all possible agent combinations and are written as vectors $\mathbf{x} = (x_1, \ldots, x_N)$ with symbols $x_i \in \{a, b, c\}$. The automorphism group of $H(N, 3)$ is composed of two groups generated by operations changing the order of elements in the vector (agent permutations) and by permutations acting on the set of symbols $\mathbf{S} = \{a, b, c\}$ (agent attributes). Namely, it is given by the direct product

$$Aut(H(N, \delta)) = \mathscr{S}_N \otimes \mathscr{S}_\delta \tag{3.12}$$

of the symmetric group $\mathscr{S}_N$ acting on the agents and the group $\mathscr{S}_\delta$ acting on the agent attributes.

Let us first look at a very small system of $N = 2$ agents and $\delta = 3$ states. The corresponding microscopic structure—the graph $H(2, 3)$—is shown on the l.h.s. of Fig. 3.7. It also illustrates the action of $\mathscr{S}_N$ on the $\mathbf{x}, \mathbf{y} \in \Sigma$, that is, the bijection induced on the configuration space by permuting the agent labels. Noteworthy, in the case of $N = 2$ there is only one alternative ordering of agents denoted here as $\hat{\sigma}_\omega(\mathbf{x})$ which takes $(x_1, x_2) \xleftrightarrow{\hat{\sigma}_\omega} (x_2, x_1)$. The respective group $\mathscr{S}_{N=2}$ therefore induces a partition in which all configurations $\mathbf{x}, \mathbf{y}$ with the same number of attributes a, b, c are *lumped* into the same set, which we may denote as $X_{\langle k_a, k_b, k_c \rangle}$. See r.h.s. of Fig. 3.7.

More generally in the case of N agents and δ agent attributes the group $\mathscr{S}_N$ induces a partition of the configuration space Σ by which all configurations with the same attribute frequencies are collected in the same macro set. Let us define $N_s(\mathbf{x})$ to be the number of agents in the configuration $\mathbf{x}$ with attribute s, $s = 1, \ldots, \delta$, and then $X_{\langle k_1, k_2, \ldots, k_\delta \rangle} \subset \Sigma$ as

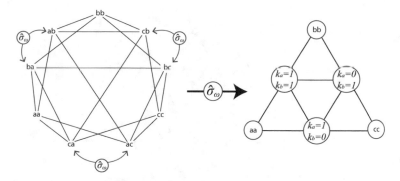

Fig. 3.7 $H(2, 3)$ and the reduction induced by $\mathscr{S}_N$

$$X_{\langle k_1,\ldots,k_s,\ldots,k_\delta \rangle} = \left\{ \mathbf{x} \in \Sigma : N_1(\mathbf{x}) = k_1, \ldots, N_s(\mathbf{x}) = k_s, \ldots \right.$$

$$\left. \ldots, N_\delta(\mathbf{x}) = k_\delta \text{ and } \sum_{s=1}^{\delta} k_s = N \right\}.$$

(3.13)

Each $X_{\langle k_1,k_2,\ldots,k_\delta \rangle}$ contains all the configurations $\mathbf{x}$ in which exactly k_s agents hold attribute s for any s. We use the notation $\langle k_1, k_2, \ldots, k_\delta \rangle$ to indicate that $\sum_{s=1}^{\delta} k_s = N$. Therefore, the reduced state space is organized as a δ simplex lattice, see Fig. 3.8.

For a model with $N = 8$ and $\delta = 3$ the resulting reduced state space is shown in Fig. 3.8. The transition structure depicted in Fig. 3.8 corresponds to the VM to which we will come back in the next chapter. The number of a, b and c agents is denoted by (respectively) k, l and m so that $\mathbf{X} = \{X_{(k,l,m)} : 0 \le k, l, m \le N, k + l + m = N\}$. The number of states for a system with N agents is $S = \sum_{i=0}^{N}(i + 1) = \frac{(N+1)(N+2)}{2}$.

For Voter-like models—used, for instance, as models of opinion and social dynamics—it is not unusual to study the dynamical behavior by looking at the time evolution of the respective attribute frequencies. It is important to notice, however, that the resulting partition is lumpable only if the transition matrix $\hat{P}$ is symmetric with respect to the action of $\mathscr{S}_N$ on Σ, namely if Theorem 3.2 holds for $\mathscr{S}_N$. The next chapter will show that this is only true for homogeneous mixing and the case of inhomogeneous interaction topologies is discussed in Chap. 7.

Let us now consider $\mathscr{S}_8$. On the l.h.s. of Fig. 3.9 the graph $H(2, 3)$ is shown along with the bijections on it induced by permutation of attributes a and c, $abc \overset{\hat{\sigma}_{81}}{\longleftrightarrow} cba$). Effectively, this corresponds to the situation of looking at "one attribute (b) against the other two ($x = a \cup c$)". Noteworthy, taking that perspective (see graph in the middle of Fig. 3.9) corresponds to a reduction of $H(2, 3)$ to $H(2, 2)$ or, more generally, of $H(N, 3)$ to the hypercube $H(N, 2)$. This means that, under the assumption of symmetric agent rules with respect to the attributes, single-step models with δ states are reducible to the binary case.

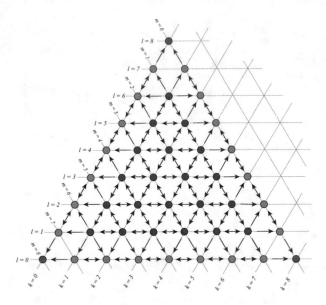

Fig. 3.8 For a three-state single step model the macroscopic process is a walk on triangular lattice (here for $N = 8$)

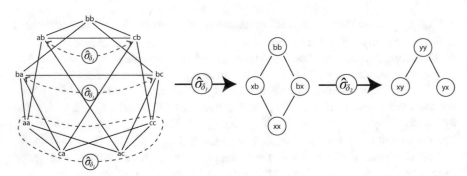

Fig. 3.9 $H(2, 3)$ and the reductions induced by $\mathscr{S}_\delta$

Moreover, even the binary case allows for further reduction (see r.h.s. of Fig. 3.9). Namely, assuming the additional symmetry $bx \xleftrightarrow{\hat{\sigma}_{\delta_2}} xb$) corresponding in a binary setting to the simultaneous flip of all agent states $x_i \to \bar{x}_i, \forall i$. The VM is a nice example in which independent of the interaction topology, $\hat{P}(\mathbf{x}, \mathbf{y}) = \hat{P}(\bar{\mathbf{x}}, \bar{\mathbf{y}})$. This reduces the state space to one half of $H(N, 2)$, which we shall denote as $H_{1/2}(N, 2)$.

The most interesting reductions can be reached by the combination of $\mathscr{S}_N$ and $\mathscr{S}_\delta$. Figure 3.10 shows possible combinations and the resulting macroscopic state spaces starting from $H(N, 3)$. For instance, partitioning $H(N, 3)$ by using the set of agent permutations $\mathscr{S}_N$ leads to state space organized as a triangular lattice (see also Fig. 3.8). Lumpability of the micro process $(\Sigma, \hat{P})$ on $H(N, 3)$ with respect to this state space rests upon the symmetry of the agent interaction probabilities with respect to all agent permutations. From the triangular structure shown on the upper right in Fig. 3.10, a further reduction ca be obtained by taking into account the symmetry of the interaction rules with respect to (at least) one pair of attributes, which we have denoted as $\hat{\sigma}_{\delta_1}$. The resulting macro process on $\mathbf{X} = (X_0, \ldots, X_N)$ is a random walk on the line with $N + 1$ states, known as Moran process for the VM interaction (after Moran 1958). In a binary setting, the macro states X_k collect all micro configurations with k agents in state $\square$ (and therefore $N - k$ agents in $\blacksquare$). Notice that a Markov projection to the Moran process is possible also for $\delta > 3$ if the micro process is symmetric with respect to permutations of (at least) $\delta - 1$ attributes. The group of transformations associated to this partition may be written as $\mathscr{S}_N \otimes \mathscr{S}_{\delta-1} \subset Aut(H(N, \delta))$.

The reduction obtained by using the full automorphism group of $H(N, 3)$ is shown on the bottom of Fig. 3.10. With respect to the Moran process on $\mathbf{X} = (X_0, \ldots, X_N)$, it means that the pairs $\{X_k, X_{(N-k)}\}$ are lumped into the same state Y_k. This can be done if we have for any k, $P(X_k, X_{k\pm1}) = P(X_{(N-k)}, X_{(N-k)\mp1})$. As a matter of fact, this description still captures the number of agents in the same state, but now information about in which state they are is omitted. This is only possible (lumpable) if the model implements completely symmetric interaction rules.

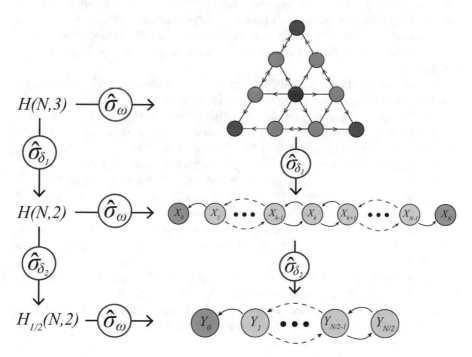

Fig. 3.10 Different levels of description are associated to different symmetry groups of $H(N, 3)$

3.5 Summary and Discussion

This chapter analyzed the probabilistic structure of a class of agent-based models (ABMs). In an ABM in which N agents can be in δ different states there are δ^N possible agent configurations and each iteration of the model takes one configuration into another one. It is therefore convenient to conceive of the agent configurations as the nodes of a huge directed graph and to link two configurations $\mathbf{x}, \mathbf{y}$ whenever the application of the ABM rules to $\mathbf{x}$ may lead to $\mathbf{y}$ in one step. If a model operates with a sequential update scheme by which one agent is chosen to update its state at a time, transitions are only allowed between system configurations that differ with respect to a single element (agent). The graph associated to those single-step models is the Hamming graph $H(N, \delta)$.

In this context, the random map representation (RMR) of a Markov process helps to understand the role devoted to the collection of (deterministic) dynamical rules used in the model from one side and of the probability distribution ω governing the sequential choice of the dynamical rule used to update the system at each time step from the other side. The importance of this probability distribution, often neglected, is to encode the design of the social structure of the exchange actions at the time of the analysis. Not only, then, are features of this probability distribution concerned with the social context the model aims to describe, but they are also crucial in predicting the properties of the macro dynamics. If we decide to remain at a Markovian level, then the partition, or equivalently the collective variables, to be used to build the model should be compatible with the symmetry of the probability distribution ω.

The fact that a single-step ABM corresponds to a random walk on a regular graph allows for a systematic study of the symmetries in the dynamical structure of an ABM. Namely, the existence of non-trivial automorphisms of the ABM micro chain tells us that certain sets of agent configurations can be interchanged without changing the probability structure of the random walk. These sets of micro states can be aggregated or lumped into a single macro state and the resulting macro-level process is still a Markov chain. If the microscopic rules are symmetric with respect agent ($\mathcal{S}_N$) and attribute ($\mathcal{S}_\delta$) permutations the full automorphism group of $H(N, \delta)$ is realized and allows for a reduction from δ^N micro to around $N/2$ macro states. Moreover, different combinations of subgroups of automorphisms and the reductions they imply are rather meaningful in terms of observables and system properties.

Notice finally that other update schemes (e.g., synchronous update) that go beyond single-step dynamics do not necessarily affect the symmetries of the micro process. The described approach may be applied to these cases as well. Extending the framework to models with continuous agent attributes is another challenging issue to be addressed by future work.

References

Banisch, S., Lima, R., & Araújo, T. (2012). Agent based models and opinion dynamics as Markov chains. *Social Networks, 34*, 549–561.

Chazottes, J.-R., & Ugalde, E. (2003). Projection of Markov measures may be Gibbsian. *Journal of Statistical Physics, 111*(5/6), 1245–1272.

Epstein, J. M. (2006). Remarks on the foundations of agent-based generative social science. In L. Tesfatsion & K. L. Judd (Eds.), *Handbook of computational economics: Agent-based computational economics* (Vol. 2, pp. 1585–1604). New York: Elsevier.

Epstein, J. M., & Axtell, R. (1996). *Growing artificial societies: Social science from the bottom up*. Washington, DC: The Brookings Institution.

Flajolet, P., & Odlyzko, A. M. (1990). Random mapping statistics. In *Advances in cryptology* (pp. 329–354). Heidelberg: Springer.

Humphreys, P. (2008). Synchronic and diachronic emergence. *Minds and Machines, 18*(4), 431–442.

Izquierdo, L. R., Izquierdo, S. S., Galán, J. M., & Santos, J. I. (2009). Techniques to understand computer simulations: Markov chain analysis. *Journal of Artificial Societies and Social Simulation, 12*(1), 6.

Kemeny, J. G., & Snell, J. L. (1976). *Finite Markov chains*. Berlin: Springer.

Levin, D. A., Peres, Y., & Wilmer, E. L. (2009). *Markov chains and mixing times*. Providence, R.I.: American Mathematical Society.

Macy, M. W., & Willer, R. (2002). From factors to actors: Computational sociology and agent-based modeling. *Annual Review of Sociology, 28*(1), 143–166.

Moran, P. A. P. (1958). Random processes in genetics. *Proceedings of the Cambridge Philosophical Society, 54*, 60–71

Squazzoni, F. (2008). The micro-macro link in social simulation. *Sociologica, 2*(1).

Chapter 4
The Voter Model with Homogeneous Mixing

This chapter is devoted to the analysis of a simple opinion model in order to illustrate the ideas developed in the previous chapter. The projection from micro to macro emphasizes the particular role played by homogeneous mixing as a requirement for the Markovianity of the projected model. We present a Markov chain analysis of the binary voter model (VM) with a particular focus on its transient dynamics and show that the general VM can be reduced to the binary case by further projection. Finally, the question of interaction constraints in form of bounded confidence is addressed. Homogeneous interaction probabilities (homogeneous mixing) are assumed in all the analyses presented in this chapter. Interaction heterogeneities are left for the next chapters.

4.1 Opinion Dynamics and Projected Systems

Voter-like models—here we shall interpret them as models of opinion and social dynamics—provide nice examples where such a projection construction is particularly meaningful. If there are δ possible agent attributes, we consider the projection Π that maps each $\mathbf{x} \in \Sigma$ into a macro state $X_{(k_1,\ldots,k_\delta)} \in \mathbf{X}$ where k_s, $s = 1,\ldots,\delta$, is the number of agents in $\mathbf{x}$ with attribute s. This captures the model dynamics in terms of frequencies of all δ attributes. The projected configuration space is then made of the $X_{(k_1,\ldots,k_\delta)}$ where $k_s \geq 0$, $s = 1,\ldots,\delta$ and $\sum_1^\delta k_s = N$. We shall now treat in detail the VM as an example of the previous ideas.

© Springer International Publishing Switzerland 2016
S. Banisch, *Markov Chain Aggregation for Agent-Based Models*,
Understanding Complex Systems, DOI 10.1007/978-3-319-24877-6_4

4.1.1   The Macro Dynamics of the Binary Voter Model

The case of a binary opinion model, $\delta = 2$, is particularly simple and therefore well-suited for an analytical starting point. In binary state models, the attribute of agent i at time t is a binary variable $x_i(t) \in \{\blacksquare, \square\} \equiv \{0, 1\}$. The opinion profile is given by the bit-string $\mathbf{x}(t) = \{x_1(t), \ldots, x_N(t)\}$ so that, as for all binary single-step dynamics, the space of all possible configurations is the set of all bit-strings of length N, $\Sigma = \{\blacksquare, \square\}^N$.

For further convenience, let us use the convention that the black state is treated as zero and the white state as one ($\blacksquare \equiv 0$ and $\square \equiv 1$) and let us define the *Hamming weight* of a configuration as

$$h(\mathbf{x}) = \sum_{i=1}^{N} x_i = N_\square(\mathbf{x}). \tag{4.1}$$

Notice that the Hamming weight is precisely the measure usually considered in the analysis of binary opinion or population genetics models as it corresponds to the opinion or gene frequency. With $N_\square = h(\mathbf{x})$ and $N_\blacksquare = N - h(\mathbf{x})$, the Hamming weight $h(\mathbf{x})$ defines the most relevant macroscopic observable ϕ of interest in the context of these models.

As stated earlier, $h(\mathbf{x})$ (just as any other macroscopic observable) defines a partition of the configuration space Σ. Namely, we can look at $h(\mathbf{x})$ as an equivalence relation such that any two configurations $\mathbf{x}, \mathbf{x}' \in \Sigma$ with $h(\mathbf{x}) = h(\mathbf{x}')$ belong to the same equivalence class. The respective equivalence classes therefore collect all configurations with the same Hamming weight, or respectively, opinion frequency. Formally, let us define $X_k \subset \Sigma$ by

$$X_k = \{\mathbf{x} : h(\mathbf{x}) = k\}. \tag{4.2}$$

Each $X_k \in \mathbf{X}, k = 0 \ldots N$ contains all the configurations ($\mathbf{x}$) in which exactly k agents hold opinion $\square$ (and then $N - k$ hold opinion $\blacksquare$). In this way we obtain a partition $\mathbf{X} = \{X_0, X_1, \ldots, X_N\}$ of the configuration space Σ. Notice that X_0 and X_N contain only one configuration, namely the final configurations $X_0 = \{(\blacksquare\blacksquare \ldots \blacksquare)\}$ and $X_N = \{(\square\square \ldots \square)\}$.

The projection of the VM micro chain $(\Sigma, \hat{P})$ yields a new macro process with state space $\mathbf{X} = (X_0, \ldots, X_N)$. This is illustrated for a small system of three agents in Fig. 4.1. Noteworthy, in this macro description, the number of states is reduced from 2^N to $N + 1$. While the number of states grows exponentially with the system size in the micro description, it grows now only linearly in the macro description.

Now, what are the conditions under which the macro process on the state space $\mathbf{X} = (X_0, \ldots, X_N)$ is again a Markov chain? It is easy to see that the partition $\mathbf{X}$ is preserved under the group of all permutations of the N agents, denoted by $\mathscr{S}_N$. Agent permutations are also compatible with the equivalence relation defined by $h(\mathbf{x})$ because $h(\mathbf{x})$ is invariant with respect to any alternative labeling of the agents.

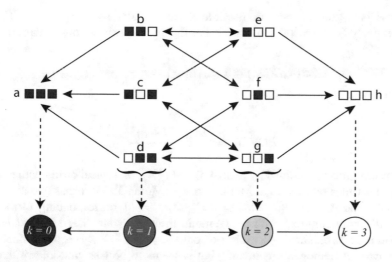

Fig. 4.1 The micro chain for the VM with three agents and its projection onto a birth-death random walk obtained by agglomeration of states with the Hamming weight $h(\mathbf{x}) = k$

We can therefore use $\mathscr{S}_N$ in the construction of a group of bijections $\mathscr{G}$ that satisfies Theorem 3.2. Namely, for each $\sigma \in \mathscr{S}_N$ we define a $\hat{\sigma} \in \mathscr{G}$ such that

$$\hat{\sigma}(\mathbf{x}) := (x_{\sigma 1}, \ldots, x_{\sigma i}, \ldots, x_{\sigma N}). \tag{4.3}$$

The respective group $\mathscr{G}$ acting on Σ preserves $\mathbf{X}$ and is compatible with $h(\mathbf{x})$: that is, for $\forall \hat{\sigma} \in \mathscr{G}$ and $\forall X_i \in \mathbf{X}$, $\mathbf{x} \in X_i$ implies that $\hat{\sigma}(\mathbf{x}) \in X_i$ and $h(\mathbf{x}) = h(\hat{\sigma}(\mathbf{x}))$. Therefore, according to Theorem 3.2, lumpability leans on the condition of the invariance of the Markov transition matrix $\hat{P}$ under the permutation group of agents, $\hat{P}(\mathbf{x}, \mathbf{y}) = \hat{P}(\hat{\sigma}(\mathbf{x}), \hat{\sigma}(\mathbf{y}))$. It is easy to see by Eq. (3.7) that this is satisfied if the probability distribution ω is invariant under the permutation group $\mathscr{S}_N$ and therefore uniform: $\omega(i, j) = 1/N^2$, for all pairs of agents (i, j).[1]

This emphasizes the particular role of homogeneous mixing in the context of these models. Homogeneous mixing is special insofar as we can get rid of the sum in Eq. (3.7) because ω is uniform. In this case $\hat{P}(\mathbf{x}, \mathbf{y})$ can be expressed only in terms of the Hamming weight of $\mathbf{x}$

$$\hat{P}(\mathbf{x}, \mathbf{y}) = \sum_{x_i \neq x_j} \omega(i, j) = \frac{h(\mathbf{x})[N - h(\mathbf{x})]}{N^2}, \tag{4.4}$$

---

[1]Notice that permutation invariance is also present if "self interactions" are excluded such that $\omega(i, i) = 0$. Then $\omega(i, j) = 1/N(N - 1)$, $\forall i \neq j$. For the following computations the possibility that an agent i "interacts" with itself is not excluded.

because only the numbers of possible attribute pairs with $(x_i, x_j) = (\square, \blacksquare)$ (respectively $(x_i, x_j) = (\blacksquare, \square)$) matter. For the macro chain we obtain therefore:

$$P(X_k, X_{k+1}) = P(X_k, X_{k-1}) = \frac{k(N-k)}{N^2}. \tag{4.5}$$

and

$$P(X_k, X_k) = \frac{k^2 + (N-k)^2}{N^2}. \tag{4.6}$$

The macro process specified by Eqs. (4.5) and (4.6) is a typical birth-death random walk sometimes referred to as Moran process (Moran 1958). It has two absorbing states X_0 and X_N for $P(X_0, X_0) = P(X_N, X_N) = 1$ corresponding to the two consensus configurations with all agents in the same state. The Markovianity of the VM process obtained by a projection onto $\mathbf{X} = (X_0, \ldots, X_N)$ is well-established in the case of homogeneous mixing, but it seems to be less well-known that the preservation of Markovianity by this projection construction is the exception rather than the rule. This is due to the fact that for heterogeneous ω the second equality in Eq. (4.4) does not hold in general, and the transition rates depend not only on the Hamming weight, but also on the population structure.

The macro chain for the binary VM with homogeneous mixing is shown in Fig. 4.2. The transition probability matrix P of the Markov chain is given by the stochastic transition matrix:

$$P = \begin{pmatrix} 1 & 0 & 0 & 0 & 0 & \cdots & 0 \\ p_1 & q_1 & p_1 & 0 & 0 & \cdots & 0 \\ 0 & p_2 & q_2 & p_2 & 0 & \cdots & 0 \\ \vdots & \ddots & \ddots & \ddots & & & \vdots \\ 0 & & p_k & q_k & p_k & & 0 \\ \vdots & & \ddots & \ddots & \ddots & & \vdots \\ 0 & \cdots & 0 & p_{N-2} & q_{N-2} & p_{N-2} & 0 \\ 0 & \cdots & 0 & 0 & p_{N-1} & q_{N-1} & p_{N-1} \\ 0 & \cdots & 0 & 0 & 0 & 0 & 1 \end{pmatrix}, \tag{4.7}$$

with $p_k = P(X_k, X_{k\pm 1})$ and $q_k = P(X_k, X_k)$ given in (4.5) and (4.6).

Fig. 4.2 Macro chain for the binary VM with homogeneous mixing

The probability that any opinion change happens in the system is $2p_k$ and then depends on the current opinion balance. But there is no general tendency of the system to be attracted by one of the extremes. In other words, the macro chain performs an unbiased random walk. Due to the particular form of p_k the prevalence of one opinion results in a reduced probability of further opinion change, contrary to the usual random walk with constant transition probabilities.

For $k \cong \frac{N}{2}$ we have $p_k \cong \frac{1}{4}$. By contrast, when k is closed to 0 or N, there is a large probability for the system to stay unchanged. Notice that for $k = 1$ or $k = N-1$ this probability tends to 1 when $N \to \infty$. This indicates that in this model once one opinion dominates over the other, public opinion as a whole becomes less dynamic, which also reveals a difficulty for new opinions to spread in the artificial society.

4.1.2 Transient Macro Dynamics

In Markov chains with absorbing states (and therefore in ABMs) the asymptotic status is quite trivial. As a result, it is the understanding of the transient that becomes the interesting issue. We shall now analyze the transient dynamics for the macro dynamics of the binary VM. In order to do so, all that is needed is to compute the fundamental matrix $\mathbf{F}$ of the Markov chain (Kemeny and Snell 1976; Behrends 2000).

Let us express P in its standard form in which the two first rows and columns stand for the absorbing states X_0 and X_N and the remaining for the $N - 1$ transient states:

$$P = \left(\frac{1 \mid 0}{R \mid Q} \right). \tag{4.8}$$

Here, Q is the $(N-1) \times (N-1)$ matrix corresponding to the transient states (without the first two rows and columns associated with X_0 and X_N). The fundamental matrix F is the inverse of $(\mathbf{1} - Q)$ where $\mathbf{1}$ is the $(N - 1) \times (N - 1)$ identity matrix. Due to the structure of P, $(\mathbf{1} - Q)$ is a tridiagonal matrix that can be inverted using, for instance, the tridiagonal matrix algorithm (also known as Thomas algorithm, Conte and Boor 1980).

For the VM, moreover, we have $P(X_k, X_{k+1}) = P(X_k, X_{k-1}) = p_k$ which allows for an analytical inversion of $(1 - Q)$. We have

$$
(1 - Q) = \begin{pmatrix}
2p_1 & -p_1 & 0 & 0 & 0 & \cdots & 0 \\
-p_2 & 2p_2 & -p_2 & 0 & 0 & \cdots & 0 \\
\vdots & \ddots & \ddots & \ddots & & & \vdots \\
0 & & -p_k & 2p_k & -p_k & & 0 \\
\vdots & & & \ddots & \ddots & \ddots & \vdots \\
0 & \cdots & 0 & 0 & -p_{N-2} & 2p_{N-2} & -p_{N-2} \\
0 & \cdots & 0 & 0 & 0 & p_{N-1} & 2p_{N-1}
\end{pmatrix}.
\tag{4.9}
$$

In order to compute $(1 - Q)^{-1}$ we can use the system of equations defined by $1 = (1 - Q)^{-1}\mathbf{F}$ and $1 = \mathbf{F}(1 - Q)^{-1}$. Due to the fact that there is only one variable per row (namely, p_k) the recursive equations that have to be solved simplify and we have as a general solution of $(1 - Q)^{-1}$:

$$
F_{ij} = \begin{Bmatrix}
\frac{i(N-j)}{Np_j} & : i \le j \\
\frac{j(N-i)}{Np_j} & : i > j
\end{Bmatrix}.
\tag{4.10}
$$

For the VM, with p_k given by Eq. (4.5) we obtain:

$$
F_{ij} = \begin{Bmatrix}
\frac{Ni}{j} & : i \le j \\
\frac{N(N-i)}{N-j} & : i > j
\end{Bmatrix}.
\tag{4.11}
$$

Equation (4.11) provides us with the fundamental matrix of the system for an arbitrary number of agents N, giving information about mean quantities of the transient dynamics in this model.

The corresponding matrix $\mathbf{G}$ that encodes information about the variance (Kemeny and Snell 1976, pp. 82–84) of the same quantities can be computed on the basis of $\mathbf{F}$ by

$$
G = \mathbf{F}(\mathbf{F}_{diag} - 1)\mathbf{F}_{square}
\tag{4.12}
$$

where $\mathbf{F}_{diag}$ contains the diagonal elements of $\mathbf{F}$ and is zero for the non-diagonal elements and $(\mathbf{F}_{square})_{ij} = (\mathbf{F})_{ij}^2$. For the VM it reads:

$$
G_{ij} = \begin{Bmatrix}
N(N - 1) & : i = j \\
(2N^2 - N)\frac{(N-i)}{(N-j)} - N^2 \frac{(N-i)^2}{(N-j)^2} & : i > j \\
(2N^2 - N)\frac{i}{j} - N^2 \frac{i^2}{j^2} & : i < j
\end{Bmatrix}.
\tag{4.13}
$$

The matrices $\mathbf{F}$ and $\mathbf{G}$ provide us with a good understanding about the transient dynamics of the VM: $F_{i,k}$ is the mean of the time the process is in the transient configuration X_k when started in the configuration X_i and $G_{i,k}$ is the corresponding variance.

An interesting quantity to characterize opinion dynamics is the time a process starting in X_k takes to end in one of the two consensual absorbing states. Defining τ_k and υ_k as the mean and the variance of the random variable for $k = 1, \ldots, N-1$ we got from (4.11) and Kemeny and Snell (1976):

$$\tau_k = N \left[\sum_{j=1}^{k-1} \frac{(N-k)}{(N-j)} + 1 + \sum_{j=k+1}^{N-1} \frac{k}{j} \right] \tag{4.14}$$

and the corresponding expression for υ can explicitly be written from (4.13) using:

$$\upsilon = (2\mathbf{F} - \mathbf{1})\tau - \tau_{sq} \tag{4.15}$$

where τ_{sq} denotes the vector resulting from τ by squaring each entry. This yields

$$\upsilon_k = 2N^2(N-k)\left[\sum_{i=1}^{k-1} \frac{1}{(N-i)} \left(\sum_{j=1}^{i-1} \frac{(N-i)}{(N-j)} + 1 + \sum_{j=i+1}^{N-1} \frac{i}{j} \right) \right] + \tag{4.16}$$

$$+(2N-1)N\left(\sum_{j=1}^{k-1} \frac{(N-k)}{(N-j)} + 1 + \sum_{j=k+1}^{N-1} \frac{k}{j} \right) +$$

$$+2N^2 k \left[\sum_{i=1}^{N-1} \frac{1}{k+i} \left(\sum_{j=1}^{k+i-1} \frac{(N-k-i)}{(N-j)} + 1 + \sum_{j=k+i+1}^{N-1} \frac{k+i}{j} \right) \right] -$$

$$-N^2 \left(\sum_{j=1}^{k-1} \frac{(N-k)}{N-j} + 1 + \sum_{j=k+1}^{N-1} \frac{k}{j} \right)^2.$$

For a system of 1000 agents, Fig. 4.3 shows the mean times until absorption τ_k from each X_k and the corresponding variances υ_k. Notice the contrast among the two scales showing how the variance is large compared with the mean.

There are interesting consequences of (4.14) and (4.16), in cases where the number of agents (N) becomes large. First, as already pointed out, we see that the ratio between the variance and the mean is quite large and in fact it diverges with N. Hence, the means are fairly unreliable estimates in this system. This is often the case for absorbing Markov chains (Kemeny and Snell 1976) making a direct interpretation of numerical simulations for this type of models tough. Even more subtle, the time scale depends significantly on the starting configuration k. In fact τ_k scales as $N \log N$ for $k = 1$ and $k = N - 1$ but as N^2 for $k = \frac{N}{2}$. We are therefore

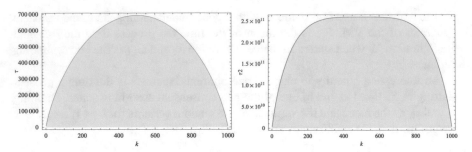

Fig. 4.3 Mean time τ_k (*l.h.s.*) and variance υ_k (*r.h.s.*) until absorption as a function of the initial configuration X_k for $N = 1000$

faced with a situation where to take the limit of asymptotic times first and then large number of agents or to do it in the reverse order is not equivalent. In other words, for a finite, even large, number of agents, there is a probability 1 of reaching one of the consensual configurations in finite time. By contrast, in the limit of an infinite number of agents this probability is 0 and the process will stay essentially in the configurations close to parity, $k = \frac{N}{2}$. Together with the presence of large fluctuations revealed in (4.16) (see Fig. 4.3) this fact is the imprint of a (dynamical) phase transition.

Besides this analysis of the scaling law of the dynamics for large N, it is also interesting to have an insight into the distributions of absorbing times for a system of fixed number of agents, the second item mentioned above. As known by the Perron-Frobenius Theorem (Seneta 2006) this distribution is exponential for large t with rate $(1 - \lambda_{max})$, λ_{max} being the maximal eigenvalue of the matrix Q. However, the correction to this distribution for intermediate times depends on the initial configuration. Indeed in our case, the distribution of the times taken by the process to fall into one of the consensual configurations departs from the exponential in a way that is strongly dependent upon the initial state, as shown in Figs. 4.4 and 4.5.

The computation of the full time distribution is based on the fact that the powers Q^t of Q contain all the information about the probability that the process is still not absorbed after t steps. To be precise, the sum over the kth row of Q^t equals the probability that the process starting at X_k is not absorbed after t iterations. This yields the cumulative distribution function shown in Fig. 4.4 for a system of 100 agents and three starting configurations $k = 1$ (green), $k = 24$ (blue) and $k = 50$ (red). The vertical dashed lines represent the respective mean values τ_k obtained using Eq. (4.14). For $k = 50$ it becomes clear that around 60% of simulation runs are absorbed until the expected absorption time is reached. Figure 4.5 shows the probability that the process is absorbed exactly at time t^{abs}. The three solid curves represent the respective probabilities for $k = 1, 24, 50$. The dashed curves are exponential functions that fit the distributions for large t^{abs} showing that the distributions decay with $(1 - \lambda_{max})$ as claimed above.

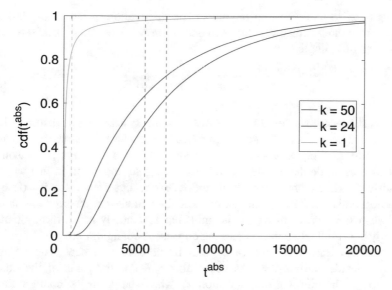

Fig. 4.4 Cumulative probability of being absorbed after time t^{abs} for $N = 100$ and three starting configurations $k = 1$ (*green*), $k = 24$ (*blue*) and $k = 50$ (*red*). *Vertical lines* show the respective expected mean convergence times τ_k

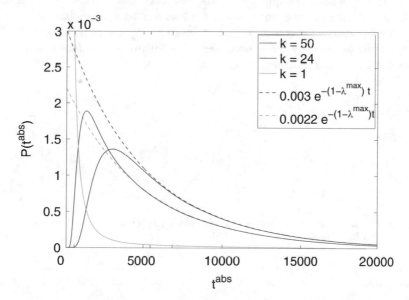

Fig. 4.5 Probability of absorbency at time t^{abs} for $N = 100$ and three starting configurations $k = 1$ (*green*), $k = 24$ (*blue*) and $k = 50$ (*red*). Exponential functions (*dashed*) are shown to illustrate the exponential decay of the convergence times as a function of $(1 - \lambda_{max})$

This leads to an interesting feature of the distribution of the absorption times coming from the fact that λ_{max} tends to one when $N \rightarrow \infty$. More precisely Seneta (2006) and (4.7) implies

$$1 > \lambda_{max} \geq 1 - p_1 \geq \frac{N-1}{N}. \tag{4.17}$$

As a consequence, we see that the times for the system to get absorbed in the final states diverge with N, and Q approaches a stochastic matrix. In fact in the limit of infinite N consensus cannot be reached. This is not the only reason why the dynamics inside the transient configurations is so important. In fact we might speculate that, in a more realistic description, exogenous events may interfere with the system and reset it from time to time, and then, in view of the previous analysis, even when the number of agents is finite but sufficiently large, the system will similarly never fall into a final absorbing consensus configuration.

Notice that Eqs. (4.11) and (4.13) can be used to gain new insight into the dynamics inside the transient. As noted above, $F_{i,k}$ is the mean of the time the process is in the transient configuration X_k when started in the configuration X_i and $G_{i,k}$ is the corresponding variance. Figures 4.6 and 4.7 show a quite different behavior depending on the initial situation. Starting from X_i close to X_1 or X_{N-1}— the strongly "biased" configurations—the residence mean times in X_k naturally decrease with the distance from i, but become almost independent of k and N for large k, whereas the corresponding variance diverges with N. Instead, starting from X_i close to $X_{N/2}$—the quasi-homogeneous configurations—the residence mean times and variance in X_k always diverge.

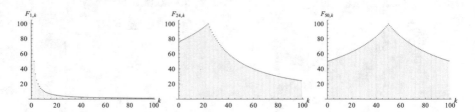

Fig. 4.6 The mean times for the process in a configuration X_k before absorption for a walk starting in X_1, X_{24} and X_{50} as function of k for $N = 100$

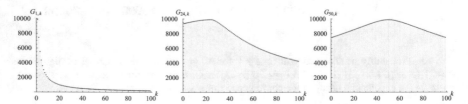

Fig. 4.7 The variance in the number of times a realization starting in X_1, X_{24} and X_{50} is in X_k before absorption as function of k for $N = 100$. Notice the scale as compared with Fig. 4.6

The reason for such "strange" behavior is quite clear: as N becomes large, almost all the realizations are trapped during very large times close to their initial configuration, see (4.7), and only very few realizations reach the opposite configurations but staying there for large times. That is, a complete overturn of the opinions is very rare but, when happened, the new situation naturally becomes as stable as the previous. Therefore we are in a case where there is almost no realization behaving as the mean. On the other hand, starting from X_i closed to $X_{N/2}$, the "homogeneous" configurations, the mean times in X_k also decrease with the distance from $N/2$, but now the mean times all scale linearly with N and the variances with N^2. Surprisingly these two behaviors—almost static on the border and very unstable "back-and-forth" in the center—compensate perfectly to end up in the same mean residence times and variance (the diagonals of $\mathbf{F}$ and $\mathbf{G}$) for all the initial configurations. The same compensation appears when we compare the probabilities for a walk stating in X_i to return in X_i, which is independent of i and almost sure for large N:

$$\lim_{t \to \infty} P^{(t)}(X_i, X_i) = \frac{F_{ii} - 1}{F_{ii}} = \frac{N - 1}{N}. \tag{4.18}$$

It is reasonable to hypothesize a correlation, if not a causal link, between fast changes in the agent opinion induced by the social process, here stylized in the dynamical rules, and the inconsistency experienced by agents between the micro and the macro level. This conflict is referred to as practical emergence (Giesen 1987). It consists of a gradual separation of the individual mental patterns from the reality. The agent is then faced with a representation that is not always perfectly in keeping with the situation (Giesen 1987, p. 342). In the opinion model, a possible rating of this practical emergence inconsistency is the mean time the macro process takes to change of state. Indeed any change of state in this process corresponds to an opinion change of an agent. Therefore the faster this rate, the smaller the switching mean time, and the more likely is the emergence of a practical disruption between picture and reality from the agent's point of view.

From (4.5) and Kemeny and Snell (1976), Theorem 3.5.6, the mean time η_k that the process remains in state X_k once the state is entered (including the entering step) is:

$$\eta_k = \frac{N^2}{2k(N - k)}. \tag{4.19}$$

Therefore, η_k is of order $\frac{N}{2}$ for k close to (but smaller than) N and 2 for k close to $\frac{N}{2}$. Again, for N large the process will be almost stationary in presence of a large majority supporting one of the opinions but extremely unstable when no opinion is clearly predominant. In the latter case practical emergence is plausible. We suggest correlating small values of η_k with this phenomenon.

Fig. 4.8 Different
realizations of simulations
with 24 out of 100 agents in
initial state □ (i.e. a process
starting in X_{24}). Markov chain
analysis shows that with
probability 0.95 the process is
in the *shaded region*

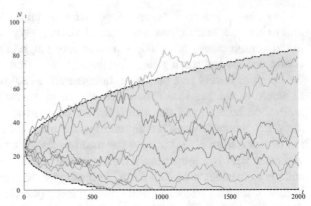

To conclude the analysis of the transient dynamics, Fig. 4.8 shows different
realizations of the agent simulation along with the expected evolution in form of
a confidence interval. The measure of the realizations inside a given confidence
interval is an increasing function of time. However, since any individual realization
may cross the border of this interval several times before falling in one of the
final absorbing states a numerical evaluation of the convergence times may be quite
delicate.

4.1.3 Exit Probabilities

The fundamental matrix **F**, Eq. (4.11), can also be used to compute the probabilities
for a process starting in X_i to end up in X_0 or X_N. These probabilities are obtained
by multiplying $\mathbf{F}R$ where R is the respective submatrix from the canonical form of
P shown in Eq. (4.8). It is well-known that the exit probabilities depend linearly on
the initial proportion of agents in the different states as

$$\lim_{t \to \infty} P^{(t)}(X_i, X_0) = \frac{N-i}{N} \qquad (4.20)$$

and

$$\lim_{t \to \infty} P^{(t)}(X_i, X_N) = \frac{i}{N}. \qquad (4.21)$$

In other words, the probability to end up in configuration (□□ . . . □) is proportional
to the initial number of □-agents.

4.1.4 Macrodynamics of the General Voter Model

For the VM with δ different attributes, the state of any agent i at time t is a variable $x_i(t) \in \{0, \dots, \delta - 1\}$. The opinion profile is given by the vector $\mathbf{x}(t) = \{x_1(t), \dots, x_N(t)\}$. The space of all possible configurations is then $\Sigma = \{0, \dots, \delta - 1\}^N$. As described in Sect. 3.2, the micro chain of single-step model with $|S| = \delta$ is a random walk with loops on the Hamming graph $H(N, \delta)$.

In the projection construction, we follow the same argument as for $\delta = 2$. We define $N_s(\mathbf{x})$ to be the number of agents in the configuration $\mathbf{x}$ with opinion s, $s = 0, \dots, \delta - 1$, and then $X_{\langle k_0, k_1, \dots, k_{\delta-1}\rangle} \subset \Sigma$ as

$$
X_{\langle k_0, \dots, k_s, \dots, k_{\delta-1}\rangle} = \left\{ \mathbf{x} \in \Sigma \; : \; N_0(\mathbf{x}) = k_0, \dots, N_s(\mathbf{x}) = k_s, \dots \right.
$$

$$
\left. \dots, N_{\delta-1}(\mathbf{x}) = k_{\delta-1} \text{ and } \sum_{s=0}^{\delta-1} k_s = N \right\}. \tag{4.22}
$$

Each $X_{\langle k_0, k_1, \dots, k_{\delta-1}\rangle}$ contains all the configurations $\mathbf{x}$ in which exactly k_s agents hold opinion s for any s. We use the notation $\langle k_0, k_1, \dots, k_{\delta-1}\rangle$ to indicate that $\sum_{s=0}^{\delta-1} k_s = N$. As in the binary case, we obtain in this way a partition $\mathbf{X}$ of the configuration space Σ.

Again, as for the binary model, the symmetry condition (3.8) of Theorem 3.2 is verified if the probability distribution ω is permutation invariant and therefore uniform: $\omega(i, j) = \frac{1}{N^2}$, for all pairs of agents (i, j). That is, the projection of the micro process $(\Sigma, \hat{P})$ onto $\mathbf{X}$ yields a Markov chain in the case of homogeneous mixing. In this case, Eq. (4.5) generalizes to:

$$
P(X_{\langle k_0, k_1, \dots, k_{\delta-1}\rangle}, X_{\langle k_0', k_1', \dots, k_{\delta-1}'\rangle}) = \frac{k_s k_r}{N^2} \tag{4.23}
$$

if $k_s' = k_s \pm 1$ and $k_r' = k_r \mp 1$ whereas $k_j' = k_j$ for all other j, and the probability that no opinion changes, Eq. (4.6), becomes

$$
P(X_{\langle k_0, k_1, \dots, k_{\delta-1}\rangle}, X_{\langle k_0, k_1, \dots, k_{\delta-1}\rangle}) = \frac{1}{N^2} \sum_{s=0}^{\delta-1} (k_s)^2.
$$

The structure of (4.23) has an interesting consequence on the dynamics of the system. We see that, if there is an s for which $k_s = 0$, the probability of transition to a state with $k_s > 0$ is zero. In other words, to change the number of agents sharing opinion s, at least one agent with such an opinion is needed. Therefore, the state space is organized as a δ-simplex with absorbing faces ordered by inclusion, corresponding to increasing sets of opinions with no supporters.

Starting in some state with no null k_s the process will finish at certain time in a state where, for the first time, $k_s = 0$ for some s (notice that only one s at each time can fall to zero since the sum of all k_s is constant). From there, the given k_s will stay equal to zero for ever, and (4.23)–(4.24) tell us that the transition probabilities are now those of a system with $\delta - 1$ opinions. Because the condition $\sum_{s=0}^{\delta-1} k_s = N$ is to be fulfilled by the remaining opinions, the system will then evolve exactly as if the N agents share $\delta - 1$ opinions from the very beginning. After a certain time a new opinion will lose all its supporters and the system is now equivalent to a full system of $\delta - 2$ opinions, and so on. The system will cascade up to the final absorbing state, with only one opinion shared by all the N agents. We recall that each of such cascade transitions is achieved in finite (random) times.

By computing the fundamental matrix of the subsystems it would be possible to access the mean and variance of the times the system evolves between two successive extinctions of group opinions. We conjecture the same scaling laws for a system of δ opinions as the ones already described for $\delta = 2$.

4.1.5 Further Reduction

Alternatively, we can make use of the symmetries in the structure of (4.23) and search for lumpable partitions to further reduce the problem. This can be done by considering the model from the perspective of a single "party" associated with (say) opinion 0. For that "party", it may be important to know how many agents are supportive because they share the same opinion, and how many are not because they support one of the remaining opinions. Thus, we reduce the model to a quasi-binary variant with the supporter opinion 0 on one side and all other opinions $(1 \cup \cdots \cup \delta - 1)$ on the other, grouping together all the states with $k_0 = r$, $r = 0, \ldots, N$.

The corresponding partition reads:

$$Y_r^0 = \bigcup_{\substack{k_1,\ldots,k_{\delta-1} \\ k_0=r}} X_{\langle r,k_1,\ldots,k_{\delta-1}\rangle}, r = 0,\ldots,N. \tag{4.24}$$

It is easy to verify that the chain (on $\mathbf{X}$) is indeed lumpable with respect to $\mathbf{Y}$ by considering the transition probabilities (4.23).[2] One can show that

$$\sum_{\substack{k_1,\ldots,k_{\delta-1} \\ k_0=r}} P(X_{\langle r,k_1,\ldots,k_{\delta-1}\rangle}, Y_{r\pm1}^0) = \frac{r(N-r)}{N^2} \tag{4.25}$$

---

[2] Alternatively, one could also verify the lumpability of $\mathbf{Y}$ directly with respect to the micro process. Namely, as shown in Sect. 3.2, the micro chain is a random walk on $H(N,\delta)$. The group $\mathscr{S}_N$ acting on the agents as well as the permutation group $\mathscr{S}_\delta$ acting on the agent attributes give rise to automorphisms of $H(N,\delta)$ such that the automorphism group is given by the direct product $Aut(H(N,\delta)) = \mathscr{S}_N \otimes \mathscr{S}_\delta$. The transformation group that generates the new partition $\mathbf{Y}$ is a subgroup of that, namely, $\Lambda = \mathscr{S}_N \otimes \mathscr{S}_{\delta-1} \subset \mathscr{S}_N \otimes \mathscr{S}_\delta$.

and therefore independent of the $k_s, s > 0$. This tells us that

$$P(Y_r^0, Y_{r+1}^0) = P(Y_r^0, Y_{r-1}^0) = \frac{r(N-r)}{N^2}. \tag{4.26}$$

It thus turns out that the chain formed by the $Y_r^0, r = 0, 1, \ldots, N$ is exactly the same as the chain derived for the binary model. Therefore, the questions regarding the evolution of one opinion in relation to all the others taken together are addressed by the transient analysis performed in Sect. 4.1.2. That is to say, from this point of view, each "party" may rely on the dynamics of a binary model as a coarse description of the evolution of its own status.

There is however an important subtlety when doing such an analysis. The asymmetry of the partition one-against-all-others will be encoded in the initial condition. For instance, starting with an equally distributed profile of N agents corresponds to the initial condition $X_{(k,k,\ldots,k)}$ in the detailed description but to $Y_{N/\delta}^0$ in the coarse case. In such a way the asymmetry in the one-against-all-others description is recovered.

In particular, this tells us that the probability to end up in the final configuration $Y_N^0 = \{(\blacksquare\blacksquare \ldots \blacksquare)\}$ is proportional to this initial bias and becomes

$$\lim_{t\to\infty} P^{(t)}(Y_{N/\delta}^0, Y_N^0) = \frac{1}{\delta}. \tag{4.27}$$

Consequently, with probability $1 - 1/\delta$ the process will transit to the class of states in which $r = 0$ with a zero probability to return to the $r > 0$ class. Viewed in the space $\mathbf{X}$, the process is not finished then, but performs a random walk until one of the uniformity states is reached, each with equal probability $1/\delta$.

4.2   Bounded Confidence and the Emergence of Opinion Polarization

An important issue in the study of opinion dynamics concerns the effects of *bounded confidence* on the model dynamics. Especially the models conceived by Hegselmann and Krause (2002) and Deffuant et al. (2001) (but also Axelrod 1997; Banisch et al. 2010) are designed to study the situation that the willingness of agents to communicate depends on the similarity of their attributes. It is also noteworthy that similarity constraints of this kind play an important role in population genetics, where they go under the label *assortative mating* (e.g., Kondrashov and Shpak 1998; Dieckmann and Doebeli 1999, see also Banisch and Araújo 2012).

In this section, we treat in detail the simplest case where bounded confidence (and other communication constraints) can be integrated, namely $\delta = 3$. Consider that agents can choose between three different alternatives $\mathbf{S} = \{a, b, c\}$. In order to model bounded confidence we define a $\mathbf{S} \times \mathbf{S}$ "confidence matrix" α which encodes for any attribute pair whether or not the attributes are compatible. If all entries in α are one, this yields the unconstrained VM with $\delta = 3$ and the results of the previous section apply. For bounded confidence, we set $\alpha(a, c) = \alpha(c, a) = 0$ meaning that the attributes a and c are incompatible ($a \nleftrightarrow c$). The consequence of this constraint is the emergence of non-consensual absorbing states, that is, the stable co-existence of different attributes.

4.2.1   The Unconstrained Case

We are particularly interested in the $\delta = 3$ case because it is the simplest version in which one can meaningfully consider the effects of bounded confidence. According to the general results of Sect. 4.1.4, the projection from micro to macro dynamics is lumpable with respect to $\mathbf{X}$ (under the homogeneous hypothesis on ω of course). We denote the number of a, b and c agents by (respectively) k, l and m so that $\mathbf{X} = \{X_{\langle k,l,m \rangle} : 0 \leq k, l, m \leq N, k + l + m = N\}$. The Markov chain topology obtained by this projection is shown in Fig. 4.9 along with the transition structure for a system of eight agents. The probabilities of the transitions are given by Eqs. (4.23) and (4.24) which allows us to compute the complete transition matrix P.

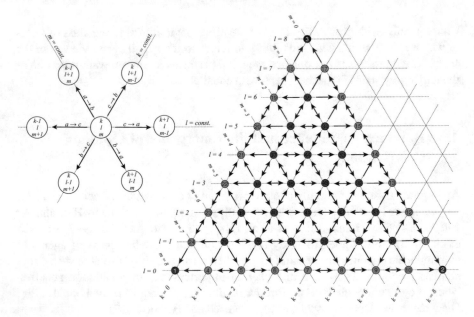

Fig. 4.9 Transition structure (*l.h.s.*) and state topology (*r.h.s.*) of the unbounded confidence model with three opinions $\mathbf{S} = \{a, b, c\}$, here $N = 8$

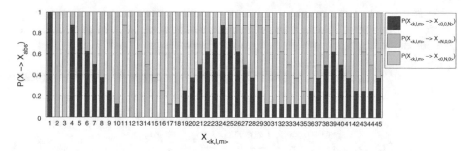

Fig. 4.10 The probabilities of reaching the three absorbing states for all initial nodes $X_{\langle k,l,m \rangle}$. Notice that all three final states can be reached only from the inner nodes (numbers 25–45)

For the construction of P, the nodes in the Markov chain are labeled in increasing order from the absorbing to the central nodes, see Fig. 4.9: labels 1–3 (black) for absorbing consensus states, labels 4–24 (blue) for two-opinion states, labels 25–39 (red) for three-opinion states with one of the opinion supporters reduced to one element, and labels 40–45 (red) for the remainder states. It is possible to compute the fundamental matrix, at least numerically if N is large, and this makes it possible to compute the significant statistical indicators of the model. For instance, if $N = 8$, the state space of the macro dynamics has 45 states and the mean times for the transient nodes to reach an absorbing state (consensus) range between 21 and 48 time steps, see Fig. 4.13. Not surprisingly the mean transition times are a function of the distance to the absorbing states as measured on the graph of the state space (Fig. 4.9).

From the fundamental matrix $\mathbf{F}$ it is also easy to compute the probabilities of ending up in each of the absorbing (consensus) states starting from any transient node using the matrix $B = \mathbf{F}R$, where R is defined as in (4.8). For instance, for $N = 8$, the absorbing probabilities for any state are represented in Fig. 4.10.

4.2.2   Bounded Confidence

Let us now turn to the question of what happens if agents with a certain opinion do not accept to change their opinion after meeting an agent of another given opinion. In the opinion dynamics literature, this is referred to as *bounded confidence* (Deffuant et al. 2001; Hegselmann and Krause 2002). From the Markov chain perspective the emergence of opinion polarization becomes a simple consequence of the restrictions posed on the interaction process. As certain transitions are excluded, the state space topology of the Markov chain changes in a way that new absorbing states become present. The respective states correspond to non-consensus configurations, hence, they represent a population with opinion clustering.

As an example, let us assume that agents in opinion state a are not willing to communicate with agents in state c and vice versa, that is to say $\alpha(a, c) = \alpha(c, a) = 0$. The corresponding Markov transition matrix P now reads:

$$P(X_{\langle k,l,m \rangle}, X_{\langle k-1,l,m+1 \rangle}) = P(X_{\langle k,l,m \rangle}, X_{\langle k+1,l,m-1 \rangle}) = 0. \tag{4.28}$$

and

$$P(X_{\langle k,l,m \rangle}, X_{\langle k,l,m \rangle}) = \left(\frac{k^2 + l^2 + m^2}{N^2} \right) + 2 \left(\frac{km}{N^2} \right). \tag{4.29}$$

The remaining entries are, as before, given by (4.23) and (4.24). The resulting state space topology is shown in Fig. 4.11, where all horizontal transition paths are removed, since those paths correspond to the $a \leftrightarrow c$ opinion changes.

For the set of bordering nodes $X_{\langle k,0,N-k \rangle} : k = 1, \ldots, N-1$ with $l = 0$ (no b-agents) there is no longer any transition that leads away from them, so that all these nodes become absorbing states. The fact that these additional absorbing states $X_{\langle k,0,N-k \rangle}$ represent opinion configurations with k agents in state a and $N-k$ agents in state c explains why the introduction of interaction restrictions leads to possible final states with opinion polarization. It is noteworthy, however, that the opinion clustering would not be observed if only one of the two transitions, $a \to c$ or $c \to a$, were excluded. In this case, there would still be a path leading away from the

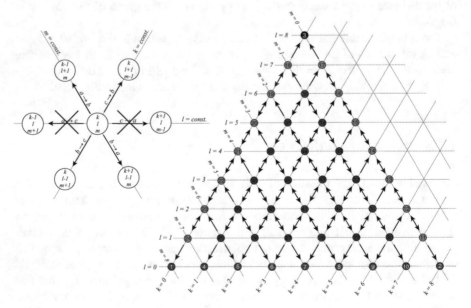

Fig. 4.11 Transition structure and state topology of the bounded confidence model for $N = 8$. All states on the $l = 0$ line (no b agents) are now absorbing states

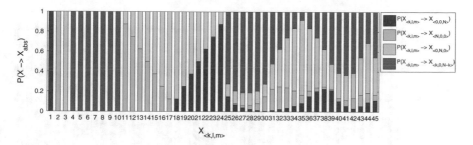

Fig. 4.12 The probabilities for all initial nodes $X_{(k,l,m)}$ converging to opinion clustering or to the three consensus nodes. Notice again that all final states and the non-consensus states in particular can be reached only from the inner nodes (numbers 25–45)

bordering nodes to one of the nodes ($X_{(0,0,N)}$ or $X_{(N,0,0)}$) in the corner of the graph. Such a set-up corresponds to an asymmetric model where the bordering atoms $X_{(k,0,N-k)} : k = 1, \ldots, N-1$ become again transient, such that the process eventually leads to the final consensus configurations as previously described. However, in the case that $a \to c$ but $c \nrightarrow a$ the final configuration $\mathbf{x} = (cc \ldots c)$ would be much more likely than $\mathbf{x} = (aa \ldots a)$, as a consequence of the asymmetry of such a model variant.

As for the unconstrained case, the fundamental matrix can be computed here as well and allows us to calculate the statistical quantities of the model such as absorbing probabilities and times. In Fig. 4.12 the probabilities of a realization starting in one of the transient states ending up in each of the absorbing final states are shown for each initial node (computed again by $B = FR$). If the process is in the first 10 nodes at $t = 0$, it will remain there forever as all these nodes are absorbing in the bounded confidence case. Notice that nothing changes for the nodes 11–24 with respect to the unconstrained case shown in Fig. 4.10. For a system in these configurations the communication constraint has no effect on the dynamics. The six absorbing non-consensus states (numbers 4–10 with only a and c opinion supporters) are reachable only from the inner nodes, that is only if all opinions are present initially. It becomes clear that for some of these configurations, the probability of converging to consensus becomes very small (e.g. nodes 25–30).

Finally, we can compare the mean time before a realization starting in a transient state remains in the transient before absorption for the bounded and the unbounded case. This statistical indicator is represented in Fig. 4.13. Notice that the times for the states 1–3 (unbounded) and 1–10 (bounded) are zero as in this case the process is absorbed from the very beginning. Again, the non-absorbing two-opinion states (11–24) are not affected.

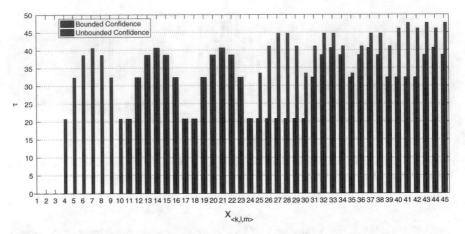

Fig. 4.13 Mean times for the transient nodes to reach an absorbing state. *Blue bars*: unbounded confidence, *red bars*: bounded confidence with $a \not\rightarrow c$. Labels of the nodes are explained in the text

4.2.3   Non-lumpability for Further Reduction

As in the general case of any δ, we can search here for lumpable partitions to further reduce the problem taking the point of view of each "party" associated with opinions a, b or c. For the unconstrained case, we have shown in Sect. 4.1.5 that the dynamics from any of these points of view reduces to the $\delta = 2$ case. The status of the bounded confidence model is different. From the perspective of opinion b the partition in "supporters" and "opponents" is lumpable, therefore, the system evolves as a binary chain (see next section). This is not the case from the perspectives of opinions a or c. For instance, from the point of view of opinion a, the corresponding partition reads:

$$Y_r^a = \bigcup_{l+m=N-r} X_{\langle r,l,m \rangle}, \quad r = 0, 1, \ldots, N. \tag{4.30}$$

and

$$P(X_{\langle r,l,m \rangle}, Y_{r+1}^a) = \frac{rl}{N^2}. \tag{4.31}$$

It turns out that the chain formed by the $Y_r^a, r = 0, 1, \ldots, N$ is not a Markov chain since the r.h.s. of (4.31) depends on l and not only on r (Kemeny and Snell 1976).

We see that the introduction of bounded confidence in this model leads to memory effects due to the fact that an agent switching from opinion a to opinion c necessarily goes through a visit to opinion b for at least one time step. Therefore, the probability of this transfer will depend on the number of supporters of opinion b at that time.

4.2.4 Transient Behavior with Bounded Confidence

As noted above, a further reduction of the Markov chain is possible if the dynamics are considered from the perspective of "party" b. The corresponding partition reads

$$Y_r^b = \bigcup_{k+m=N-r} X_{\langle k,r,m\rangle}, \quad r = 0, 1, \ldots, N \tag{4.32}$$

and the transition probabilities are

$$P(X_{\langle k,r,m\rangle}, Y_{r+1}^b) = \frac{r(N-r)}{N^2}. \tag{4.33}$$

As discussed in Sect. 4.1.5, the probability to converge to $Y_N^b = \{(bb\ldots b)\}$ is $1/\delta = 1/3$ and the probability to end up in one of the configurations in Y_0^b is $2/3$, provided that the model is initialized with an equal number of a, b and c agents. Notice that contrary to the unbounded confidence case the process is really in its final state whenever $r = 0$ as all configurations in Y_0^b are absorbing.

The convergence times (see Fig. 4.13 for a small system) are composed of the (relatively short) time needed to end up in the class of states Y_0^b including those of opinion polarization and the (relatively long) time needed to converge to Y_N^b. In the following, we use a transformation proposed in Kemeny and Snell (1976, pp. 64/65), in order to assess the two times independently. Notice that all the results obtained in this section are also applicable to the binary VM (with $X_k \equiv Y_k^b$) to study the effects of initial opinion bias.

The basic idea is to "compute all probabilities relative to the hypothesis that the process ends up in the given absorbing state" (Kemeny and Snell 1976, p. 64). This leads to a new absorbing chain with the specified state as the single absorbing state. In fact, for our purpose, it is not necessary to completely determine the transition matrix for that new chain as the fundamental matrix of the original process ($\mathbf{F}$) can be used directly to compute the fundamental matrix of the new chain ($\tilde{\mathbf{F}}$). Let $B = \mathbf{F}R$ where $\mathbf{F}$ is the fundamental matrix of the binary chain (4.11) and R the respective $2 \times N$ submatrix of the canonical form (4.8). The elements b_{1j} (b_{2j}) of B correspond to the exit probabilities of the process started in j to end up in $X_0 \equiv Y_0^b$ ($X_N \equiv Y_N^b$). Recall that $b_{1j} = (N-j)/N$ and $b_{2j} = j/N$ (see Sect. 4.1.3). Now let D_0 be a diagonal matrix with $d_{jj} = b_{1j}$ and respectively define D_N as $d_{jj} = b_{2j}$. Then, according to Kemeny and Snell (1976, p. 65), the fundamental matrices of the new chains with absorbing state $X_0 \equiv Y_0^b$ and $X_N \equiv Y_N^b$ respectively is given by

$$\tilde{\mathbf{F}}_0 = D_0^{-1}\mathbf{F}D_0$$

$$\tilde{\mathbf{F}}_N = D_N^{-1}\mathbf{F}D_N. \tag{4.34}$$

In our case with the fundamental matrix given in Eq. (4.11) we obtain

$$(\tilde{\mathbf{F}}_0)_{ij} = b_{1j}\frac{F_{ij}}{b_{1i}} = \begin{cases} \frac{iN(N-j)}{j(N-i)} & : i \leq j \\ N & : i > j \end{cases} \tag{4.35}$$

and

$$(\tilde{\mathbf{F}}_N)_{ij} = b_{2j}\frac{F_{ij}}{b_{2i}} = \begin{cases} N & : i \leq j \\ \frac{jN(N-i)}{i(N-j)} & : i > j \end{cases}. \tag{4.36}$$

The fundamental matrices $\tilde{\mathbf{F}}_0$ and $\tilde{\mathbf{F}}_N$ allow for a very good understanding of the average behavior of the model. $\tilde{\mathbf{F}}_0$ encodes the mean number of steps that the realizations which eventually converge to Y_0^b pass through any state Y_r^b, and $\tilde{\mathbf{F}}_N$ informs us about the mean behavior of realizations that end up in Y_N^b. For instance, we can compute the mean convergence time to each absorbing state independently. For convergence to Y_0^b from the initial state Y_r^b we have

$$\tilde{\tau}_0(r) = rN + \sum_{j=r+1}^{N} \frac{rN(N-j)}{j(N-r)} \tag{4.37}$$

and for convergence to uniformity corresponding to Y_N^b

$$\tilde{\tau}_N(r) = (N-r)N + \sum_{j=1}^{r-1} \frac{jN(N-r)}{r(N-j)} \tag{4.38}$$

For a system of 100 agents these times are shown in Fig. 4.14. It becomes clear that the mean convergence times to Y_0^b and Y_N^b are equal if the initial situation is unbiased, that is, if there are $r = N/2$ agents with attribute b and $N/2$ agents in the other two states a or c. However, with an increasing initial bias, there is an increasing gap between average convergence time to one or the other absorbing state. For the system with three possible attributes a, b and c and random initial conditions the initial number of b agents is around $N/3 \approx 33$. This is illustrated by the dashed vertical line. In that case, the mean convergence time for realizations that end up in possible polarized configurations with only a and c agents becomes considerably smaller compared to the configuration with all agents in state b.

$\tilde{\mathbf{F}}_0$ and $\tilde{\mathbf{F}}_N$ enable moreover to study the transient behavior of the model with initial bias in more detail. For a system of 100 agents, Fig. 4.15 shows the mean number of steps that a process ending up in Y_0^b (l.h.s.) and Y_N^b (r.h.s.) is in the transient states provided the model is initialized with 33 agents in b and 67 agents in a or c. This information is encoded in the 33rd row of $\tilde{\mathbf{F}}_0$ and $\tilde{\mathbf{F}}_N$ respectively. We first comment on the l.h.s. showing the mean behavior of realizations ending

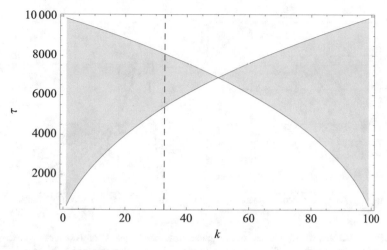

Fig. 4.14 Mean convergence times to $X_0 \equiv Y_0^b$ (*red*) and $X_N \equiv Y_N^b$ (*blue*) independently. The *vertical dashed line* represents the initial bias for the model with $\delta = 3$

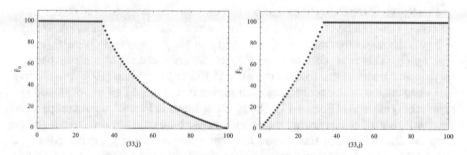

Fig. 4.15 Mean number of steps a process that eventually converges to Y_0^b (*l.h.s.*) and Y_N^b (*r.h.s.*) is in the transient states for a system of 100 agents and an initial number of $r = 33$ agent in state b

up in an absorbing configuration where only a and c agents remain (Y_0^b). The figure shows that, in average, all the transient states that are closer to Y_0^b than the initial configuration in Y_{33}^b are met N times. Naturally, the states "to the right" are encountered less frequently. It should be clear that the entries of $\tilde{\mathbf{F}}_0$ should not be read as the mean behavior of every single realization, but rather as the average behavior over a large series of realizations. For instance, the mean number of steps to Y_{99}^b, which is very close to the opposite absorbing state, is approximately $1/2$. However, this does not mean that every second realization approaches the opposite absorbing state so closely. It rather means that there are rare realizations that take that way, and once they are at the opposite extreme, these realizations have a high chance to stay there for some while. In fact, the fundamental matrix $\tilde{\mathbf{F}}_0$ tells us that, once a realization reached Y_{99}^b, the mean number of returns to that state is $N - 1$. The interpretation of the r.h.s. ($\tilde{\mathbf{F}}_N$) goes in the same way.

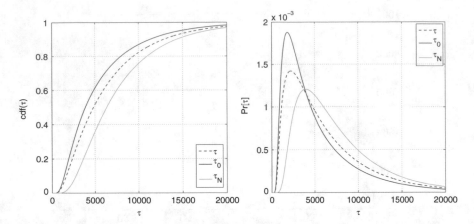

Fig. 4.16 Probability distribution of convergence times for a system of 100 agents when started with $r = 33$. Convergence to Y_0^b, τ_0, is considerably faster than convergence to Y_N^b, τ_N

Finally, the probability distribution of convergence times to one or the other absorbing state can be computed easily for a given N. In this computation, we first compute the respective matrices $\tilde{Q}_0$ as $(\tilde{Q}_0)_{ij} = (b_{1j}/b_{1i})Q_{ij}$ and $\tilde{Q}_N$ as $(\tilde{Q}_N)_{ij} = (b_{2j}/b_{2i})Q_{ij}$. This is in complete analogy to the computation of the independent fundamental matrices and follows the work of Kemeny and Snell (1976, pp. 64–65). The computation of the probability distribution is then based on the evaluation of powers of $\tilde{Q}$ as done in Sect. 4.1.2. The result is shown in Fig. 4.16. For comparison, the distribution of convergence times to either absorbing state (dashed, red) is shown for $r = 33$. All in all, this shows how the general convergence behavior is a composite of the two different convergence trends obtained by considering the two absorbing states independently.

4.3   Simple Generalizations

We first mention an easy generalization of the existence of absorbing states for the case of bounded confidence in a model with any number δ of different opinions. In order to get non-consensual absorbing states it is necessary and sufficient that a subset of opinions is mutually incommunicable. In this case all the states belonging to the simplex generated by the mutually incommunicable opinions become absorbing. It is worthwhile noticing that absorbing states may appear in different clusters of simplexes provided that the corresponding opinions are related by chains of communicating links. An example of this type appears for $\delta = 3$ if $(a \nleftrightarrow b)$ and $(a \nleftrightarrow c)$ but $(b \leftrightarrow c)$ where the absorbing states are either the simplex with only a and b or with only a and c opinions.

Another interesting issue concerns agent models with vectorial (or equivalently matrix or table) individual attribute space. Suppose that at each time step each agent i is characterized by a list of q attributes, where the first attribute may take n_1 possible values, the second attribute n_2 values and so on up to the qth attribute with n_q possible values. The corresponding ABM can then be easily built as in Sect. 4.1.4 by taking $\delta = n_1 \times n_2 \times \cdots \times n_q$. As long as one is interested in following the macrodynamics obtained by lumping all agent configurations with an equal attribute frequency for all the δ attributes, the reduction proposed in Sect. 4.1.5 also applies. Moreover, absorbing non-consensual states will appear in exactly the same way as described above as a consequence of bounded confidence.

For this vectorial opinion model there is, however, an unexpected subtlety when we are interested in the macrodynamics of the agents ranked by only one of their attributes, for instance, if the agents are separated in n_1 different groups according to the number of agents sharing their first attribute. Then, the partition is no longer lumpable, and therefore the evolution of the corresponding random variables (for instance, the number of elements of each group) is not a Markov chain. Again, in this case, new memory effects may appear from this choice of aggregation to build the macrodynamics. The proof can be done as in (4.30) and (4.31).

4.4 Summary and Discussion

In this chapter, we considered the VM from a Markovian perspective and derive explicit statements about the possibility of linking a microscopic agent model to the dynamical processes of macroscopic observables that are useful for a precise understanding of the model dynamics. In this way the dynamics of collective variables (i.e., opinion frequency) may be studied, and a description of macro dynamics as emergent properties of micro dynamics, in particularly during transient times, is possible.

Using Markov chain computations, we obtain a very detailed understanding of the VM with homogeneous mixing. On the one hand, the computation of the fundamental matrix of the macro chain provides us with precise knowledge about the mean transient behavior, on the other, it also tells us that some care must be taken in order to relate those mean quantities to single realizations of the model. Regarding convergence times, full information (probability distribution of convergence times) is provided by numerical integration over the transient states which gives a better idea of the transient behaviors that single realizations may exhibit. The analysis is extended to the general (multi-state) VM, the analysis of which is reducible to the binary case in the absence of interaction constraints. On the other hand, similarity constraints as bounded confidence or assortative mating lead to additional absorbing states in the macro chain. This shows that opinion polarization is a direct consequence of bounded confidence (see Chap. 8 for a biological interpretation in terms of sympatric speciation).

This is what makes homogeneous mixing (and respectively, the complete graph) so special because the full permutation invariance is realized $(Aut(K_N) = \mathscr{S}_N)$. On the other hand, an important mark of ABMs is their ability to include arbitrary levels of heterogeneity and stochasticity (or uncertainty) into the description of a system of interacting agents. In a sense, the partition of the configuration space defining the macro level of the description has to be refined in order to account for an increased level of heterogeneity or a falloff in the symmetry of the probability distribution. It is, however, clear that, in absence of any symmetry, there is no other choice for this partition than to stay at the micro level and, in this sense, no Markovian description of a macro level is possible in this case. This will be spelled out in detail in the next chapter.

References

Axelrod, R. (1997). The dissemination of culture: A model with local convergence and global polarization. *The Journal of Conflict Resolution, 41*(2), 203–226.

Banisch, S., & Araújo, T. (2012). Who replaces whom? Local versus non-local replacement in social and evolutionary dynamics. *Discontinuity, Nonlinearity, and Complexity, 2*(1), 57–73.

Banisch, S., Araujo, T., & Louçã, J. (2010). Opinion dynamics and communication networks. *Advances in Complex Systems, 13*(1), 95–111.

Behrends, E. (2000). *Introduction to Markov chains with special emphasis on rapid mixing.* Berlin: Friedrick Vieweg & Son.

Conte, S. D., & Boor, C. W. D. (1980). *Elementary numerical analysis: An algorithmic approach* (3rd ed.). New York: McGraw-Hill Higher Education.

Deffuant, G., Neau, D., Amblard, F., & Weisbuch, G. (2001). Mixing beliefs among interacting agents. *Advances in Complex Systems, 3*, 87–98.

Dieckmann, U., & Doebeli, M. (1999). On the origin of species by sympatric speciation. *Nature, 400*(6742), 354–357.

Giesen, B. (1987). Beyond reductionism: Four models relating micro and macro levels. In: J. C. Alexander, B. Giesen, R. Münch, & N. J. Smelser (Eds.), *The micro-macro link.* Berkeley, CA: University of California Press.

Hegselmann, R., & Krause, U. (2002). Opinion dynamics and bounded confidence: Models, analysis and simulation. *Journal of Artificial Societies and Social Simulation, 5*(3), 1.

Kemeny, J. G., & Snell, J. L. (1976). *Finite Markov chains.* Berlin: Springer.

Kondrashov, A. S., & Shpak, M. (1998). On the origin of species by means of assortative mating. *Proceedings of the Royal Society of London Series B, 265*, 2273–2278.

Moran, P. A. P. (1958). Random processes in genetics. *Proceedings of the Cambridge Philosophical Society, 54*, 60–71.

Seneta, E. (2006). *Non-negative matrices and Markov chains.* Springer series in statistics (2nd ed.). New York: Springer.

Chapter 5
From Network Symmetries to Markov Projections

In the third chapter, we have seen that an agent-based model (ABM) defines a process of change at the individual level—a micro process—by which in each time step one configuration of individuals is transformed into another configuration. For a class of models we have shown this micro process to be a Markov chain on the space of all possible agent configurations. Moreover, we have shown that the full aggregation—that is, the re-formulation of the model by mere aggregation over the individual attributes of all agents—may give rise to a new process that is again a Markov chain, however, only under the rather restrictive assumption of homogeneous mixing. Heterogeneities in the micro description, in general, destroy the Markov property of the macro process obtained by such a *full aggregation*.

The question addressed in this chapter is how to derive Markovian coarse-grainings (Markov projections) if the assumption of homogeneous mixing is relaxed. In other words, how must the micro model and the projection construction be so that the projected system is still a Markov chain? We develop a tool which relates symmetries in the interaction topology to partitions of the configuration space with respect to which the micro process is lumpable. In effect, this leads to a refinement of the full aggregation which exploits all the dynamical redundancies that have its source in the agent network on which the model is implemented. Notably, the result is stated in terms of the symmetries of the agent network which is much simpler than the micro chain on the configuration space where the aggregation process (lump) is achieved. Some of the results presented here are also available under Banisch and Lima (2013).

© Springer International Publishing Switzerland 2016
S. Banisch, *Markov Chain Aggregation for Agent-Based Models*,
Understanding Complex Systems, DOI 10.1007/978-3-319-24877-6_5

5.1   Interaction Heterogeneity and Projection Refinement

Let us begin this chapter with the simple example that is running through this thesis. Consider the VM with three agents on different networks defined by a 3×3 adjacency matrix A with $a_{ij} = 1$ whenever i and j are connected. As before, in the iteration process, an agent pair (i,j) is chosen at random out of the set of all agent pairs with $a_{ij} = 1$ and the first adopts the state of the second. Notice that an alternative way of realizing the agent update is to first choose an agent i and then choose another agent j out of all agents connected to i. The former is called *link update* and the latter *node update* dynamics and we shall see that this can lead to different probability distributions ω. We mainly consider link update in this chapter, but comment on the differences between the two variants in Sect. 5.2.

We first consider the complete graph defined by $a_{ij} = 1$ whenever $i \neq j$ and $a_{ii} = 0$. Notice that in that case, the two update variants are lead to the same $\omega(i,j)$. Namely, the probability of a pair (i,j) to be chosen is for every pair $\omega = 1/6$. That is, except for the exclusion of self-choice (with $\omega(i,i) = 0$) it leads to the case dealt with in the previous chapter. Figure 5.1 briefly recalls the respective Markov chain formulation and projection by illustrating (1) the connectivity structure $\omega(i,j) = \omega; \forall i \neq j$, (2) the micro chain this leads to along with the transition rates, and (3) the resulting macro chain.

In order to go beyond complete homogeneity let us consider what happens to that picture of one link is removed. Therefore, let us assume that $a_{23} = a_{32} = 0$. Under link update this leads to the following interaction probabilities: $\omega(1,2) = \omega(2,1) = \omega(1,3) = \omega(3,1) = \omega = 1/4$, and $\omega(2,3) = \omega(3,2) = 0$. This topology, the resulting micro chain and the probabilistic effects on the macro level are shown in Fig. 5.2.

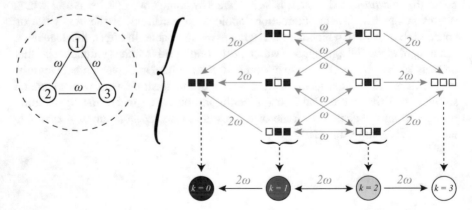

Fig. 5.1 Probabilistic structure of the model with three agents on the complete graph

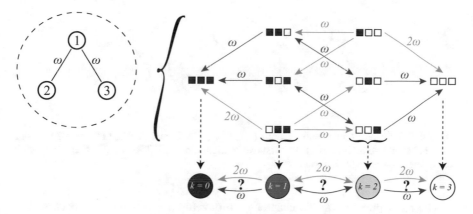

Fig. 5.2 Probabilistic structure of the model with three agents if the connection between 2 and 3 is absent

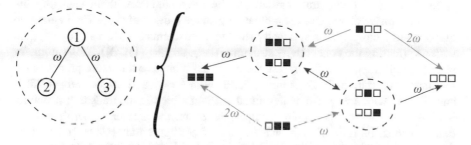

Fig. 5.3 Refinement of the partition that preserves Markovianity

It becomes clear that the introduction of interaction heterogeneity translates into irregularities in the probabilistic structure of the micro chain in a way that the symmetry condition in Theorem 3.2, $\hat{P}(\mathbf{x}, \mathbf{y}) = \hat{P}(\hat{\sigma}(\mathbf{x}), \hat{\sigma}(\mathbf{y}))$, is violated for the macro partition $\mathbf{X} = (X_0, X_1, X_2, X_3)$. In other words, it leads to the non-lumpability of the partition $\mathbf{X} = (X_0, X_1, X_2, X_3)$. As shown in Fig. 5.2 the transition probabilities at the macro level are not uniquely defined and depend upon the respective micro configuration. Consider, as an example, the transitions from X_2 to X_3. The probability (3.7) of a transition form configuration (■□□) to (□□□) is $\hat{P}(\blacksquare\square\square, \square\square\square) = \omega(1, 2) + \omega(1, 3) = 2\omega$, whereas $\hat{P}(\square\blacksquare\square, \square\square\square) = \omega(2, 1) + \omega(2, 3) = \omega$ and $\hat{P}(\square\square\blacksquare, \square\square\square) = \omega(3, 1) + \omega(3, 2) = \omega$. While all these probabilities are equal for the complete graph (as $\omega(i, j) = \omega : \forall i, j$) they are not all equal if one or two connections are absent which violates the lumpability condition.

Deriving a partition such that the micro process projected onto it is a Markov chain requires a refinement of the aggregation procedure. For the example considered here the respective refined partition is shown in Fig. 5.3.

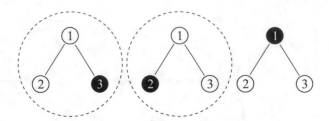

Fig. 5.4 The three different configurations (□□■), (□■□) and (■□□) of length 3 with one agent in ■ and two in □ ($k = 2$). The first two configurations (□□■) and (□■□) are what we will call macroscopically equivalent

The main purpose of this chapter is to develop a systematic approach to this projection refinement by exploiting all the dynamical redundancies resulting from the symmetries of agent network. Network symmetries can be used to identify bundles of micro configurations that can be interchanged without changing the hypercubic micro chain. Our example may provide a first intuition. The interaction graph in our example has a symmetry such that the agents 2 and 3 can be permuted without affecting the connectivity structure, (i.e., $Aut_\omega = (1)(23)$). This symmetry imposes a symmetries in the hypercube graph associated to the micro process such that the configurations (□■□) and (□□■) with $k = 2$ and respectively (■■□) and (■□■) with $k = 1$ can be permuted without affecting the transition structure. See also Fig. 5.4. In this simple example, therefore, the previous macro atoms X_2 (and X_1) must be refined such that the sets of configurations {(■■□), (■□■)} (respectively {(□■□), (□□■)}) on the one hand and {(□■■)} (respectively {(■□□)}) on the other form different sets in a Markovian partition.

5.2   Social Structure at the Micro Level

The effect of different social networks on the dynamics of ABMs plays an increasingly important role in the research of these models. Certain aspects of the model behavior may sometimes be very different when implemented on different topologies. One might be surprised that in the consideration of the micro level dynamics (Sect. 3.2, previous chapter) nothing is said about how different agent networks incorporate into this framework. The simple reason is that the role of networks in the models is essentially to determine the interaction probabilities of agents and that we consider that kind of information via the probability distribution ω.

For instance, in the VM two agents (i, j) linked in the network are chosen at random. From the network it is possible to infer directly the respective probability $\omega(i, j)$. As mentioned earlier, there are two different ways of agent choice: first, one can first choose an agent i and then choose another agent j out of its neighborhood (node update dynamics); second, both agents are chosen at one instance by the choice of an edge in the network (link update dynamics). In general, the two modes of agent choice lead to a different $\omega(i,j)$. Node update leads to

$$\omega(i,j) = \frac{1}{N}\frac{1}{k_i}, \tag{5.1}$$

where k_i is the degree of agent i. For the second version with link choice, on a graph with adjacency matrix A the probability $\omega(i,j)$ is

$$\omega(i,j) = \frac{a_{ij}}{|E|}, \tag{5.2}$$

where a_{ij} is the element in A corresponding to the edge (i,j) and $|E|$ is the total number of edges in the graph.

Figure 5.5 illustrates the differences in the probability distribution ω for node and link update using the small example considered in the previous section. Notice that in any case $\sum \omega(i,j) = 1$ for it is a distribution over agent choices. Notice moreover, that the symmetry (1)(23) is preserves for the two update schemes.

The notion of ω is quite general and allows also to incorporate other types of social structure. There may be cases in which agents are heterogeneous with respect to certain static characteristics; for instance, if they belong to different ethnical groups or working classes. This might effect not only there likeliness to meet in the model, but also the choice probabilities for different behavioral options, that is, the behavior of agents within their group may be different from the agent behavior across different groups. All those effect are encoded into the probability distribution ω along with the social network of agents.

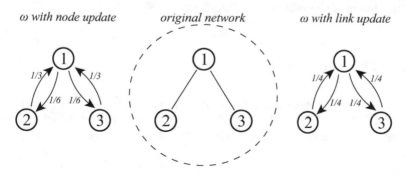

Fig. 5.5 Node versus link update in the example considered above

5.3   Markovian Aggregation

For those reasons, it is convenient to formulate the theoretical ideas presented in this section in terms of the probability distribution ω. We first define the notion of macroscopic equivalence, then we consider the VM and finally we discuss the generalization to the class of models with single-step dynamics (see Sect. 3.2.3).

5.3.1   Macroscopic Equivalence

Let $(\Sigma, \hat{P})$ be a micro chain corresponding to an ABM. Let $\mathcal{M}$ be a partition of the configuration space Σ with respect to which the micro process is lumpable. For further convenience we define the following notion of macroscopic equivalence:

Definition 5.1 Two configurations $\mathbf{x}$ and $\mathbf{x}'$ are macroscopically equivalent if the lumpability condition (Kemeny and Snell 1976, Theorem 6.3.2)

$$\hat{p}_{\mathbf{x}Y} = \sum_{\mathbf{y} \in Y} \hat{P}(\mathbf{x}, \mathbf{y}) = \sum_{\mathbf{y} \in Y} \hat{P}(\mathbf{x}', \mathbf{y}) = \hat{p}_{\mathbf{x}'Y} \tag{5.3}$$

is satisfied for all $Y \in \mathcal{M}$. Then $\mathbf{x}$ and $\mathbf{x}'$ belong to the same atom X of the partition $\mathcal{M}$.

The notion of macroscopic equivalence is motivated by the fact that two macroscopically equivalent configurations contribute in exactly the same way to the dynamical behavior of the macro process on $\mathcal{M}$. It is important to notice that macroscopic equivalence is inherently linked to a partition $\mathcal{M}$, that is, with a macro description of the process, because two configurations that are equivalent with respect to one partition might not be with respect to another.

5.3.2   The Voter Model

Let $Aut_\omega(N)$ be the subgroup of the permutations σ acting on the set $\mathbf{N}$ of agents such that $\omega(\sigma i, \sigma j) = \omega(i, j)$ for all $i, j \in \mathbf{N}$. To each $\sigma \in Aut_\omega(N)$ we associate a $\hat{\sigma}$ which is a bijection on the configuration space Σ. If $\mathbf{x} \in \Sigma$ with $\mathbf{x} = (x_1, \ldots, x_i, \ldots, x_N)$ then

$$\hat{\sigma}(\mathbf{x}) = (x_{\sigma 1}, \ldots, x_{\sigma i}, \ldots, x_{\sigma N}). \tag{5.4}$$

We now define a partition $\mathcal{M}_\omega$ of Σ using $Aut_\omega(N)$. Two configurations $\mathbf{x}, \mathbf{x}' \in \Sigma$ belong to the same atom of the partition $\mathcal{M}_\omega$ iff there is a $\sigma \in Aut_\omega(N)$ such that $\mathbf{x}' = \hat{\sigma}(\mathbf{x})$. Clearly this is an equivalence relation and therefore it defines a partition on Σ.

Proposition 5.1 *The partition $\mathcal{M}_\omega$ is lumpable for the agent model on Σ with agent choice based on ω and therefore the corresponding projected process is a Markov chain.*

Proof Consider the distribution of interaction probabilities ω and its permutation group of symmetries $Aut_\omega(N) = \{\sigma : \omega(\sigma i, \sigma j) = \omega(i, j), \forall i, j \in \mathbf{N}\}$. Suppose we know (at least) one configuration (the generator) $\mathbf{x}^k \in \Sigma$ for each $X_k \subset \Sigma$ and construct the partition $\mathcal{M}_\omega = (X_1, \ldots, X_k, \ldots)$ by

$$X_k = Aut_\omega(N) \circ \mathbf{x}^k = \bigcup_{\forall \hat{\sigma}} \hat{\sigma}(\mathbf{x}^k). \tag{5.5}$$

A necessary and sufficient condition for lumpability is that the transition probability from a configuration $\mathbf{x} \in X_k$ to any atom $X_s \in \mathcal{M}_\omega$ be the same for all $\mathbf{x} \in X_k$ (Kemeny and Snell 1976, Theorem 6.3.2). That is, we have to show macroscopic equivalence, Eq. (5.1), for the pairs of configurations $\mathbf{x}$ and $\hat{\sigma}(\mathbf{x})$. By Theorem 3.2 we know that this is satisfied whenever $\hat{P}(\mathbf{x}, \mathbf{y}) = \hat{P}(\hat{\sigma}(\mathbf{x}), \hat{\sigma}(\mathbf{y}))$ for any $\sigma \in Aut_\omega(N)$.

In the VM, for the case that $x \overset{i}{\sim} y$ we know that $x_j = y_j$ for all j except i and that the transition requires the choice of an edge $(i, .)$. Denoting $x_i = s$ and $y_i = \bar{s}$ we rewrite Eq. (3.7) as

$$\hat{P}(\mathbf{x}, \mathbf{y}) = \sum_{j:(x_j=\bar{s})} \omega(i, j). \tag{5.6}$$

If $\mathbf{x} \overset{i}{\sim} \mathbf{y}$ it is easy to show that $\hat{\sigma}(\mathbf{x}) \overset{\sigma i}{\sim} \hat{\sigma}(\mathbf{y})$ and we know that $s = \hat{\sigma}(x_{\sigma i}) \neq \hat{\sigma}(y_{\sigma i}) = \bar{s}$. The transition therefore requires the choice of an edge $(\sigma i, .)$. We obtain

$$\hat{P}(\hat{\sigma}(\mathbf{x}), \hat{\sigma}(\mathbf{y})) = \sum_{k:(\sigma(x_k)=\bar{s})} \omega(\sigma i, k). \tag{5.7}$$

Given an arbitrary configuration $\mathbf{x}$, for any j with $x_j = \bar{s}$ we have a corresponding $k = \sigma j$ with $\hat{\sigma}(x_k) = \bar{s}$ because $x_j = \bar{s} \Leftrightarrow \hat{\sigma}(x_{\sigma j}) = \bar{s}$. That is, the summations in Eqs. (5.6) and (5.7) are equal for any σ for which $\omega(i, j) = \omega(\sigma i, \sigma j)$. This is true by the definition of $Aut_\omega(N)$ for all permutations $\sigma \in Aut_\omega(N)$.

5.3.3 Single-Step Dynamics

Proposition 5.1 can be applied without modification to any interacting particle system in which the local transition probabilities are a function solely of the local neighborhood configuration, as defined by an unchanging graph.[1] For the class of models with single-step dynamics the proof can be done following the same argument.

As discussed in Sect. 3.2.3, the update from one time step to the next is defined by a function $\mathbf{u} : \mathbf{S}^r \times \Lambda \rightarrow \mathbf{S}$ that depends on the attributes of an arbitrary number of agents (r) and on an additional variable $\lambda \in \Lambda$ accounting for a possible stochastic part in the update mechanism. The probability distribution ω is therefore over $r + 1$-tuples $(\omega(i, j, \ldots, k, \lambda))$. For the construction of a partition $\mathcal{M}_\omega$ we must now consider the group of $\sigma \in Aut_\omega$ with $\omega(i, j, \ldots, k, \lambda) = \omega(\sigma i, \sigma j, \ldots, \sigma k, \sigma \lambda)$. Then, as before, classes of macroscopically equivalent configurations (and therewith $\mathcal{M}_\omega$) are defined by $\mathbf{x}' = \hat{\sigma}(\mathbf{x})$ with $\hat{\sigma}(\mathbf{x})$ as in (5.4).

5.4 The Two-Community Model

5.4.1 Model

Consider a population composed of two sub-population of size L and M such that $L + M = N$ and assume that individuals within the same sub-population are connected by strong ties whereas only weak ties connect individuals that belong to different communities. We could think of that in terms of a spatial topology with the paradigmatic example of two villages with intensive interaction among people of the same village and some contact across the villages. This is similar to the most common interpretation in population genetics where this is called the island model (Wright 1943). In another reading the model could by related to status homophily (Lazarsfeld and Merton 1954) accounting for a situation where agents belonging to the same class (social class, race, religious community) interact more intensively than people belonging to different classes (Fig. 5.6).

Let us adopt the perspective of a weighted graph and say that an edge with weight $a_{ij} = 1$ connects agents of the same community whereas edges across the two communities have a weight $a_{ij} = r$. Therefore, r is the ratio between strong and weak ties. For the VM run on such a network, notice again that there may be subtle differences in the resulting interaction probabilities $\omega(i, j)$ depending on how the agent choice is performed. First, in the case of link update dynamics a link (i, j) is chosen out of the set of all links and so the $\omega(i, j)$ are proportional to the edge weight. Namely, let γ denote the interaction probability between agents of the same

---

[1] I am grateful to an anonymous reviewer for this formulation.

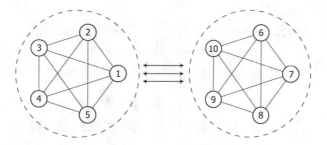

Fig. 5.6 A two-component graph with two homogeneous sub-populations

community and α the respective probability across communities, then

$$\gamma = \frac{1}{2LM + ((L-1)L + (M-1)M)r} \tag{5.8}$$

$$\alpha = \frac{r}{2LM + ((L-1)L + (M-1)M)r}, \tag{5.9}$$

where the divisor is the sum over all edge weights and establishes that $\sum_{(i,j)} \omega(i,j) = 1$. A second mode of agent choice is to first choose an agent i and then choose a second agent j out of its neighbor set. In the case that $M \neq L$, the interaction probabilities become different from (5.9). In the following, however, we will concentrate on the example with $M = L = 50$, and in this case Eq. (5.9) gives the right interaction probabilities for node and link update dynamics.

5.4.2 Markov Projection

Notice, moreover, that independent of M and L both update modes give rise to the same symmetry group $Aut_\omega(N) = (1 \ldots M)(M+1 \ldots N)$. $Aut_\omega(N)$ is composed of the symmetric group $\mathscr{S}_L$ and $\mathscr{S}_M$ acting on the two subgraphs and it means that ω is invariant under permutations of agents within the same community.[2] Let us denote by m and l the number of $\square$-agents in M and L. It is then clear that all configurations $\mathbf{x}$ and $\mathbf{y}$ with $[m(\mathbf{x}) = m(\mathbf{y})] \cap [l(\mathbf{x}) = l(\mathbf{y})]$ are macroscopically equivalent. As $0 \leq m \leq M$ and $0 \leq l \leq L$ the aggregation defines a Markov chain with $(M+1)(L+1)$ states which is still very small compared to the number of $2^{(M+L)}$ micro configurations. Notice that this generalizes naturally to a larger number of subgraphs. Notice also that the multipartite graphs studied in Sood and

---

[2]Notice that the case $M = L$ is special because it leads to additional symmetries as the two communities are interchangeable. This is not generally the case and therefore we develop the more general case of $M \neq L$ here, even if the computations are mostly performed for the example $M = L = 50$.

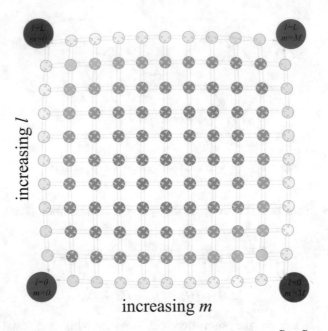

increasing l

increasing m

Fig. 5.7 The structure of the chain for $L = M = 10$. The consensus states $\tilde{X}_{0,0}, \tilde{X}_{M,L}$ as well as the states of inter-community polarization $\tilde{X}_{0,L}, \tilde{X}_{M,0}$ are *highlighted*. The quasi-stationary distribution is mapped into node colors from *blue* (low values) to *red* (high values)

Redner (2005) fall into this category and that the authors used the respective sub-densities in their mean-field description.

The structure of the Markov chain associated to the VM on the two-community graph is shown in Fig. 5.7. For the system of size M and L the transition probabilities for the transitions leaving an atom $\tilde{X}_{m,l}$ are given by

$$P(\tilde{X}_{m,l}, \tilde{X}_{m+1,l}) = \gamma(m(M-m)) + \alpha(M-m)l$$

$$P(\tilde{X}_{m,l}, \tilde{X}_{m-1,l}) = \gamma(m(M-m)) + \alpha m(L-l)$$

$$P(\tilde{X}_{m,l}, \tilde{X}_{m,l+1}) = \gamma(L-l)l + \alpha(L-l)m$$

$$P(\tilde{X}_{m,l}, \tilde{X}_{m,l-1}) = \gamma(L-l)l + \alpha(M-m)l$$

The four states located at the corners are highlighted in Fig. 5.7. The atom on the lower left ($l = 0, m = 0$) and the upper right corner ($l = L, m = M$) correspond to the two states of complete consensus where all agent in the two communities have adopted the same state. These are the absorbing states of the process. The other two ($l = L, m = 0$ and $l = 0, m = M$) correspond to the situation that agents within the same community are aligned, but there is disagreement between the different communities. This is a form of "local alignment and global polarization", and especially if the coupling across the communities becomes weak, there is a relatively high transient (quasi-stationary) probability for those situations. We will refer to them as *inter-community polarization*.

In what follows, we study a system with $M = L = 50$. This gives a Markov chain of size $(M + 1)(L + 1) = 2601$. Notice that the computations (matrix inversion and powers) needed in the analysis of that chain bear already some computational cost and that a further increase in system size will increase these costs greatly.

5.4.3 Convergence Times

We start the analysis of the model behavior on the two-community topology by computing the mean number of steps required to reach a final consensus configuration $\tilde{X}_{0,0}$ or $\tilde{X}_{M,L}$. The mean convergence times can be computed on the basis of the fundamental matrix $\mathbf{F}$ which contains the mean number of visits before absorption for all node pairs (see Sect. 4.1.2). Figure 5.8 compares the mean convergence times for all initial states $\tilde{X}_{m,l}$ and a coupling ratio of $r = 1/100$ (l.h.s.) to the homogeneous mixing situation with $r = 1$ (r.h.s.). In comparison to the homogeneous mixing case [Eq. (4.14)] the mean number of steps before absorption increases considerably for all initial configurations. For $m + l = k = 50$ the complete graph will order in average after 6880 steps whereas for a weak coupling with $r = 1/100$ this number increases to 9437 for the completely disordered configurations with $m = 25, l = 25$. Notably, it increases further to 11921 for the initial configurations with consensus in the communities but disagreement between the two islands (polarization).

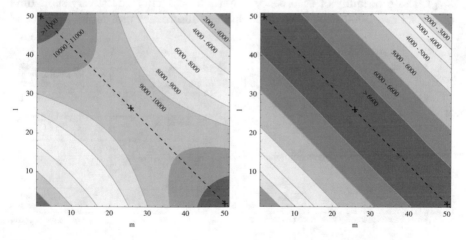

Fig. 5.8 Mean convergence times for $M = L = 50$ for all initial configurations $X_{m,l}$. The disordered initial configuration $X_{M/2,L/2}$ and the two ordered configurations $\tilde{X}_{M,0}$ and $\tilde{X}_{0,L}$ are *highlighted* by +. On the *l.h.s.* $r = \alpha/\gamma = 1/100$, on the *r.h.s.* $r = 1, \alpha = \gamma$

Notice, that in the homogeneous situation where $\alpha = \gamma$ the convergence times are only a function of the total density $k = m + l$. For every two atoms $\tilde{X}_{m_1,l_1}, \tilde{X}_{m_2,l_2}$ for which $m_1 + l_1 = m_2 + l_2$ we obtain the same mean convergence time. This is not surprising, of course, because for $\alpha = \gamma$ the Markov chain on $\tilde{\mathbf{X}}$ is lumpable with respect to the full aggregation $\mathbf{X} = (X_0, \ldots, X_k, \ldots, X_N)$ and so the behavior of every $\tilde{X}_{m,l}$ within the same X_k is identical, from the macro perspective. This changes if the VM is run on the two-community topology and the coupling between the communities is smaller than the coupling among agents in the same island (i.e., $\alpha < \gamma$). Due to the topological effects and a reduced communication across communities, the emergence of a meta-stable configuration of local alignment within communities but global polarization across communities is likely. In general, the process requires more time to converge if initialized in a situation where one community is ordered from the beginning and the opinions diverging from that consensus concentrate in the other community. Notably, an increase in convergence times is observed even for the initial configurations which are completely disordered (e.g., $\tilde{X}_{25,25}$), because a considerable number of realizations is first driven to a state of inter-community polarization before it eventually evolves further to a consensus profile (see below).

We compare these two situations (namely initial disorder $\tilde{X}_{25,25}$ and initial order $\tilde{X}_{50,0}$) by considering the distribution of convergence times for two configurations with $m + l = N/2 = 50$. The respective cumulative distributions for $r = 1/100$ is shown on the l.h.s. of Fig. 5.9 and on the r.h.s. the respective probability of absorbency at time t is shown.

In the case of initial disorder (red curve), where the states $\square$ and $\blacksquare$ are distributed equally over the two islands, there is a certain number of realizations that approaches one absorbing consensus state without entering the states of partial order ($\tilde{X}_{M,0}$ and $\tilde{X}_{0,L}$). The probability of absorbency reaches a peak after a relatively short time of around $t \approx 3000$ steps whereas the highest absorbency probability lies around

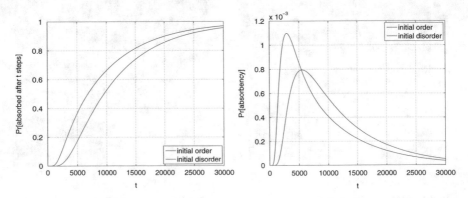

Fig. 5.9 Distribution of convergence times τ for $M = L = 50$, $r = 1/100$ for the disordered initial configuration $\tilde{X}_{M/2,L/2}$ and the two ordered configurations $\tilde{X}_{M,0}$ and $\tilde{X}_{0,L}$ with inter-community polarization

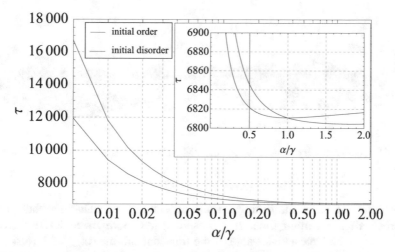

Fig. 5.10 Mean convergence times τ as a function of the relative inter-group interaction strength $r = \alpha/\gamma$ for the ordered initial configuration $\tilde{X}_{M,0}$ and the disordered initial configuration $\tilde{X}_{M/2,L/2}$

$t \approx 5000$ for the ordered initial condition. At around $t \approx 5000$ already 40 % of realizations have converged for the disordered case, but only 20 % in case of initial order. This shows that there is a strong influence of the interaction topology leading to a high heterogeneity between different initial configurations with the same global magnetization $k = m + l$. The ordered configurations $\tilde{X}_{M,0}$ and $\tilde{X}_{0,L}$ function as dynamical traps and it may take a long time to escape from them especially when r becomes small. On the other hand, however, Markov chain theory tells us that the probability for very long waiting times decays exponentially.

In Fig. 5.10, a more detailed picture of how convergence times increases as $r = \alpha/\gamma$ decreases is provided. For the two initial situations considered previously the mean convergence times are shown as a function of $r = \alpha/\gamma$. Notice again that these extreme configuration are highlighted by $+$ in Fig. 5.8. It is clear that the mean times to absorbency diverge as r approaches zero, $\lim_{r \to 0} \tau = \infty$. This is due to the fact that the interaction topology becomes disconnected in that extreme case, and therefore, the non-consensus configurations $\tilde{X}_{M,0}$ and $\tilde{X}_{0,L}$ become absorbing. In other words, to go from (say) $\tilde{X}_{0,L}$ to (say) $\tilde{X}_{0,0}$ requires an infinite number of steps. In fact, we then deal with a completely new chain that has four absorbing states, or more precisely, with two chains one for each island. However, as long as $r > 0$ the possibility to escape from $\tilde{X}_{0,L}$ remains, even if it takes very long.

5.4.4 Quasi-Stationary Distribution

Finally, to characterize the long-term transient behavior, let us look at the quasi-stationary distribution of the two-community VM. This distribution contains the

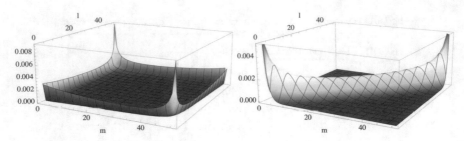

Fig. 5.11 Quasi-stationary distribution for the VM on two islands with $r = 1/100$ (*l.h.s.*) and $r = 1$ (*r.h.s.*)

probabilities to be in the different transient states for realizations that are not absorbed after a certain time. It corresponds the normalized left eigenvector associated to the largest eigenvalue of the transient sub-matrix Q of P (just as the stationary distribution of a regular chain is the normalized left eigenvector of the transition matrix P). See, for instance, Darroch and Seneta (1965) (pages 91–93 in particular) for a description of the quasi-stationary distribution.

Figure 5.11 shows the quasi-stationary distribution for the two-community VM with $r = 1/100$ and $r = 1$. Notice again that the later corresponds to the homogeneous mixing case. If r is small there is a high (conditional) probability that the process is trapped in one of the states of inter-community polarization. Also the states $\tilde{X}_{m,0}$ and $\tilde{X}_{0,l}$ with one uniform sub-population have a relatively high probability indicating that convergence to complete consensus out of local order does not happen via a transition through complete disorder. This is in stark contrast to the homogeneous mixing situation, which is shown on the r.h.s. of Fig. 5.11. In this case, states of inter-community polarization ($m = M, l = 0$ and $m = 0, l = L$) and states close to that become in effect very rare random events.[3]

5.5   On the Role of Peer-to-Peer Communication
in an Opinion Model with Leader

This section presents a Markov chain treatment of the VM on a topology that models leadership. We show how the probability that the leader imposes its opinion on a follower population increases with the influence asymmetry between a leader and the followers and is independent of peer-to-peer processes among followers.

---

[3]The reason for this is clear. The number of micro configurations $x \in \Sigma$ mapped into the state $\tilde{X}_{m,l}$ is $\binom{M}{m}\binom{L}{l}$ which is a huge number for $m \approx M/2, l \approx L/2$ but only 1 for $m = M, l = 0$ and $m = 0, l = L$. Because under homogeneous mixing there is no favoring of particular agent configurations with the same $k = m + l$ the stationary probability at macro scale is proportional to the cardinality of the set $\tilde{X}_{m,l}$.

A greater influence does not only increase the respective exit probability, it also accelerates the convergence process. However, this acceleration is undermined by a stronger peer-to-peer communication among followers.

5.5.1   Model

Here we study the binary VM on an asymmetric topology. Namely, we introduce an opinion leader that has an increased influence on the rest of the agent population. Therefore, consider a homogeneous population of N agents indexed by $i = 1, \ldots, N$. As before, each agent can adopt two opinions, $\square$ or $\blacksquare$. Consider further that there is another agent (the leader indexed by $i = 0$) with a stronger influence on the population such that the probability that its attribute spreads in the population is increased. In principle, we also want to allow that the population influences the leader, but the probability of such an event is rather small.

The VM operates by choosing an agent (i) which adopts the opinion of one of its neighbors (j). As before, $\omega(i, j)$ is the probability that the pair (i, j) is chosen. Under link update, and with the convention that the first agent (i) imitates the second (j), leadership can be included by introducing an asymmetry in the interaction probabilities such that it is more probable to choose the leader in second place. Then, it becomes more probable that its state is adopted by another agent. We index the leader with a 0 and assign the following interaction probabilities:

$$\omega(0, j) = \beta,$$
$$\omega(i, 0) = \alpha,$$
$$\omega(i, j) = \gamma, \tag{5.10}$$

$i, j = 1, \ldots, N$. An increased probability that the leader opinion is adopted by a follower is modeled by $\alpha \gg \beta$. The third probability, γ, accounts for the probability of a peer-to-peer interaction which does not involve the leader. Notice that with Eq. (5.10) the model is formulated directly in terms of interaction probabilities ω by which we avoid a possible confusion between link and node update dynamics (Sect. 5.2). The resulting interaction topology is shown in Fig. 5.12.

5.5.2   Markov Projection

It is clear that ω is still highly symmetric, because $\omega(i, j) = \gamma$ (5.10) for all agents in the follower population. More precisely, ω is invariant under all permutations $\mathscr{S}_N$ of the N agents in the follower population and therefore the topology shown in Fig. 5.12 is topologically equivalent to the star graph of size $N + 1$ (in fact, with

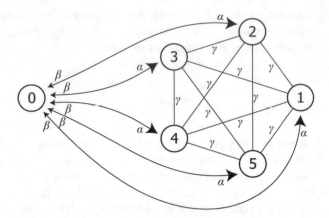

Fig. 5.12 Agent interaction topology in which one agent (indexed by 0) has an increased influence on the rest of the population

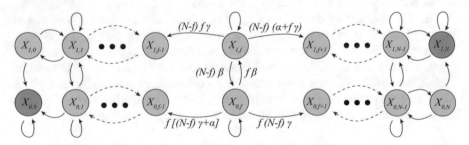

Fig. 5.13 The macro chain associated with the binary VM on the leader-follower topology. The transition probabilities are also shown

$\gamma = 0$ it is a star). Notice, moreover, that the leader topology is actually a special case of the two-community structure obtained by setting $M = 1$ and $L = N$.

Namely, the topology shown in Fig. 5.12 gives rise to the symmetry group $Aut_\omega = (0)(1 \ldots N)$. In this case, a lumpable partition is obtained by the independent observation of the state of the leader x_0 and the number of follower agents in the different states (given by $f = N_\square(\mathbf{x}/x_0)$). This partition is given by $\tilde{\mathbf{X}} = \{\tilde{X}_{l,f} : l = 0, 1; f = 0, \ldots, N\}$ with

$$\tilde{X}_{l,f} = \{\mathbf{x} : N_\square(x_0) = l \cap N_\square(\mathbf{x}/x_0) = f\}. \tag{5.11}$$

In words, the subsets $\tilde{X}_{0,f}$ contain all configurations $\mathbf{x} \in \Sigma$ in which the leader is in state ■ and f follower agents are in state □. Respectively, the subsets $\tilde{X}_{1,f}$ contain the configurations with $x_0 = \square$ and f follower agents in state □. It is thus clear that for the leader-follower system with total size of $N + 1$ the macro chain has $2(N + 1)$ states. The associated chain along with the corresponding transition probabilities is shown in Fig. 5.13.

In principle, all the information about the convergence probabilities and the transient behavior can be obtained by computing the fundamental matrix of that system (Kemeny and Snell 1976, Chap. 3). For the leader-follower system the transition matrix P is a $2(N+1) \times 2(N+1)$ dimensional matrix and the matrices Q and $\mathbf{F} = (\mathbf{1} - Q)^{-1}$ are of size $2N \times 2N$. Hence, the computation of the fundamental matrix requires the inversion of a matrix of that size. At least numerically, this is doable on an up-to-date computer platform for systems of several thousand agents.

5.5.3 Simple Observations

However, let us begin the analysis with two simple observations. First of all, it is clear from Fig. 5.13 and the transition probabilities that the macro chain has two absorbing states corresponding to the uniformity configurations $\tilde{X}_{1,N} \equiv (\square\square \ldots \square)$ and $\tilde{X}_{0,0} \equiv (\blacksquare\blacksquare \ldots \blacksquare)$.

Secondly, notice that in case $\beta = 0$—the case that followers have no influence on the leader at all—there is a zero probability that the leader changes its state. This means that the macro chain is disconnected and the model performs a random walk on the upper or the lower chain depending on the initial state of the leader (see Fig. 5.13). It is clear, then, that the process converges to the configuration in which all follower agents adopted the state of the leader.

5.5.4 Influence of the Leader

How does an increasing asymmetry ($\alpha > \beta$) effect the model behavior? To address this question we compute the probability that the system converges to the initial state of the leader. Consider the opinion leader is in state $\square$ at $t = 0$. The probability that all agents end up in $\square$ as $t \to \infty$ is

$$Pr_{t \to \infty}[\tilde{X}_{1,f} \to \tilde{X}_{1,N}] = \frac{\alpha + f\beta}{\alpha + N\beta}. \tag{5.12}$$

As the ratio $\frac{\alpha}{\beta}$ increases, the chances that the system converges to the leader's opinion increase very fast in the beginning, and approach 1 in the limit $\frac{\alpha}{\beta} \to \infty$. Notice that even in the case that all followers are against the opinion of the leader, an influence ratio of $\frac{\alpha}{\beta} > N$ is sufficient to obtain a chance of $Pr_{t \to \infty}[\tilde{X}_{1,0} \to \tilde{X}_{1,N}] > 1/2$ that the leader imposes its opinion against the consensus opinion in the follower population. More precisely, for $\frac{\alpha}{\beta} = N$, Eq. (5.12) becomes $Pr_{t \to \infty}[\tilde{X}_{1,0} \to \tilde{X}_{1,N}] = \frac{1}{2} + \frac{f}{N}$. If the leader has such a strong influence, convergence to its state hence becomes the most probable option.

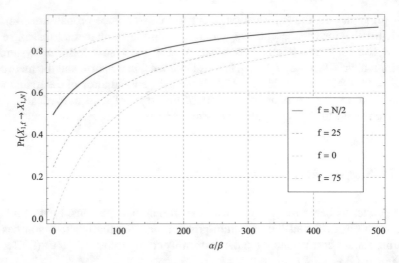

Fig. 5.14 The probability that the leader imposes its opinion onto the entire population as a function of $\frac{\alpha}{\beta}$ for different initial proportions f of follower agents in the leader's state

Figure 5.14 shows this probability as a function of the ratio $\frac{\alpha}{\beta}$ for different initial proportions f of follower agents in the same state as the leader. We set $N = 100$ in Eq. (5.12) in which case $f = N/2 = 50$ corresponds to the case that the followers are divided into two groups of equal size.

It is noteworthy, that the exit probabilities do not depend on the strength of the peer-to-peer interaction γ. Therefore the influence of the opinion leader is as if there was no communication at all among the followers. At a first sight this seems a bit counter-intuitive, but considering that γ does not introduce any bias in favor of one or the other consensus state it is not too surprising (see Fig. 5.13).

Finally, we see from Eq. (5.12) that the case $\alpha = \beta$ restores the results obtained in the VM with homogeneous mixing (and in general for the VM on undirected networks). For the configurations with the leader in $\square$ we obtain for $Pr_{t\rightarrow\infty}[\tilde{X}_{1,0} \rightarrow \tilde{X}_{1,N}] = \frac{f+1}{N+1}$ where $f + 1$ is just the total number of individuals in state $\square$ and $N + 1$ the total number of agents.

5.5.5   Convergence Times

We now look at the mean convergence times as a function of the network parameters α, β and γ for a finite system of $N = 100$ followers and one leader. Because ω is a distribution over all agent pairs, we have $N[\alpha + \beta + (N - 1)\gamma] = 1$. This means that there are effectively two free parameters in that analysis and it is convenient to study the network influence in term of the ratios α/β and γ/β. To obtain the mean convergence time, the fundamental matrix is computed for different

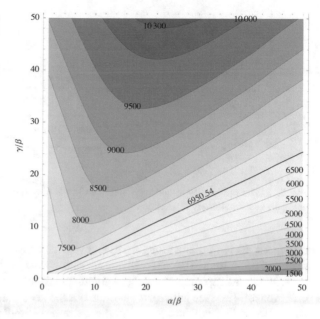

Fig. 5.15 Mean convergence times for $\alpha/\beta = 1 \ldots 50$ and $\gamma/\beta = 1 \ldots 50$. The *thick blue line* indicates the convergence time for the VM with homogeneous mixing

ratios $\alpha/\beta = 1, 2, \ldots 50$ and $\gamma/\beta = 1, 2, \ldots 50$ from which the respective mean convergence times can be obtained directly (Kemeny and Snell 1976, pp. 49–51). Notice that with any relative increase of α/β and γ/β the probability β decreases and this means that a leader change becomes less likely. Increasing α/β corresponds to an increasing asymmetry between leader and followers, an increase in γ/β to an intensification of the (symmetric) mutual influence in the follower population.

The result of this analysis is shown in form of a contour plot in Fig. 5.15. We notice two basic opposing trends in that plot. First, a stronger influence of the leader (increasing α/β) tremendously speeds up the convergence to one of the absorbing consensus states. The leader strongly drives the entire system towards its initial state and with a relatively high probability consensus is reached without a state change of the leader. On the other hand, however, an increasing mutual influence among the followers (increasing γ/β) may rule out this effect and slow down the process so that convergence to a final consensus state takes more time. Noteworthy, in a large area of the parameter space it takes even more time than the VM with homogeneous mixing of the same size (obtained with $\alpha = \beta = \gamma$). This is illustrated by the thicker blue contour line. For all parameter configurations above this line, mean convergence times become larger compared to the homogeneous mixing case.

Let us consider two examples. First, the case $\alpha/\beta = 100$ and $\gamma/\beta = 1$ in a system of $N = 100$ followers with an initial number of $N/2 = 50$ followers and the leader in □. With this parameter constellation, the mean number of steps until convergence is (only) 887 steps. It is easy to compute that in this relatively short period, approximately 60 % of the realizations have been absorbed, 50 % in $\tilde{X}_{1,N}$ and 10 % in $\tilde{X}_{0,0}$ (the latter involving at least one change of the leader). Virtually all

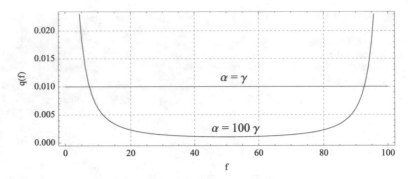

Fig. 5.16 The probability to observe f followers in the long run for the two examples computed using the quasi-stationary distribution

remaining 40 % of realizations are in fact very close to absorbency: in $\approx 33.5\,\%$ of the cases more than 90 % of followers are in the state of the leader.

While a stronger mutual influence among the followers does not affect the overall exit probabilities (see above), it may rule out the acceleration of convergence to the state a strong leader. Consider, as another extreme example, the case $\alpha/\beta = 100$ and $\gamma/\beta = 100$ in a system of $N = 100$ followers with an initial number of $N/2 = 50$ followers and the leader in $\square$. This leads to an average convergence time of 9278 steps and to a long-term transient behavior in which all follower constellations are equally likely (see below, Fig. 5.16).

5.5.6   Transient Behavior

In order to obtain a complete picture of the transient behavior of the model, we first compute the quasi-stationary distribution for the two examples. The respective probability to observe f followers in state $\square$ in the long run is shown in Fig. 5.16. The blue curve represents the case of a strong leader ($\alpha/\beta = 100$ and $\gamma/\beta = 1$). Equation (5.10) tells us that in this case, an agent pair $(i, 0)$ (follower, leader) will be chosen in half of the cases (as $N\alpha = 1/2$), the probability for the choice of two follower agents is also close to one half, namely $N(N - 1)\gamma = 99/200$. Therefore, a constellation $(0, j)$ (leader, follower) is chosen in average only once in 200 time steps. Notice, moreover, that a state change really takes place only if $x_0 \neq x_i$ and that therefore the change of the leader becomes even more unlikely because the leader has already imposed its opinion on most of the followers. Therefore, even if the parameter constellation allows that followers change the state of the leader ($\beta > 0$), a persistent situation in which the follower population opposes the leader or at least remains close to the fifty-fifty configuration cannot be observed. Once the leader changes its state, it quite immediately drives the population of followers to the opposite extreme corresponding to its opinion.

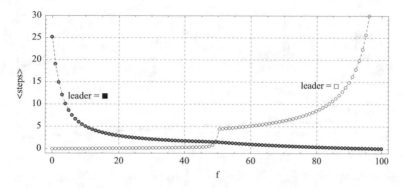

Fig. 5.17 Mean number of steps a process starting in $\tilde{X}_{1,50}$ is in the different atoms $\tilde{X}_{1,f}$ (*light*) and $\tilde{X}_{0,f}$ (*dark*) for $\alpha/\beta = 100$ and $\gamma/\beta = 1$

The most likely behavior of the model with strong leader is also highlighted by the mean hitting times shown in Fig. 5.17. It shows the expected number of steps a process starting in $\tilde{X}_{1,50}$ is in the different atoms $\tilde{X}_{1,f}$ (white circles for $x_0 = \square$) and $\tilde{X}_{0,f}$ (dark circles for $x_0 = \blacksquare$). Notice that the initial state $\tilde{X}_{1,50}$ corresponds to the light circle at $f = 50$. In order to approach the more probable absorbing state $\tilde{X}_{1,N}$ the process has to transit through all the states to the right with $f > 50$, and consequently the mean hitting times of these states are high compared to the rest. If, on that way, a state change of the leader takes place, the process goes to the other extreme passing through the transient states with intermediate f rather rapidly. Finally, the drop off in the hitting time to the left of the initial state gives an idea of how strong the leader influence is in this case. There is virtually a zero probability that social influence processes among the followers drive the system far from the leader state and, in fact, at least in a small system of 100 followers a second state change of the leader is quite rare.

In the second example with $\alpha/\beta = 100$ and $\gamma/\beta = 100$ the model behaves in a different way. First of all, the quasi-stationary distribution shown in Fig. 5.16 (red curve) tells us that, in the long run, all follower constellations are equally likely. This decoupling from the leader is surprising if we recall that the exit probabilities are strongly biased in favor of the initial state of the leader (see Eq. (5.12) and Fig. 5.12) and are not affected by the strong peer-to-peer interaction.

The mean hitting times for that example are shown in Fig. 5.18. It becomes clear that the behavior strongly resembles the behavior of the VM with homogeneous mixing (see Fig. 4.6, previous chapter). But despite these similarities in mixing behavior the exit probabilities remain strongly biased for $\alpha/\beta = 100$. The main reason for the difference in the transient behavior is that the effective influence rate of the leader is reduced tremendously as γ approaches α. By Eq. (5.10) we see that the probability of choosing an agent pair $(i, 0)$ (follower, leader) becomes in fact

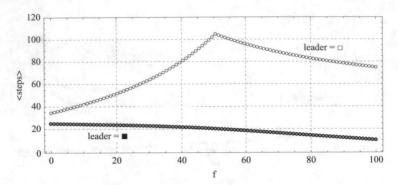

Fig. 5.18 Mean number of steps a process starting in $\tilde{X}_{1,50}$ is in the different atoms $\tilde{X}_{1,f}$ (*light*) and $\tilde{X}_{0,f}$ (*dark*) for $\alpha/\beta = 100$ and $\gamma/\beta = 100$

very small now (as $N\alpha = 100/10001 \approx 1\,\%$), the probability for the choice of two follower agents instead effectively increases to $N(N-1)\gamma = 9900/10001 \approx 98.99\,\%$. Consequently, a constellation $(0,j)$ (leader, follower) in which the leader could change its state is chosen in average only once in 10,000 time steps.

5.5.7 Alternative Interpretation

In fact, the second example with $\alpha/\beta = 100$ and $\gamma/\beta = 100$ (and consequently $\alpha = \gamma$) calls for an interpretation not in terms of leader and followers, but suggests to understand the "leader" as a member of the group that is just not so amenable to influence, compared to the others. Namely, $\omega(i,0) = \alpha = \gamma = \omega(i,j)$ means that followers give the same importance to the leader as to any other follower agent. On the other hand, β becomes very small so that the chances that followers change the leader are reduced. Of course, such a situation may appear only in a small system, but not at the scale of populations. It may happen at the group level of (say) 20 individuals (of course, only in an approximate sense). In this context, the analysis shows that a single stubborn individual can strongly influence the outcome of a consensus process (in small groups).

5.6   The Ring

5.6.1 Strongly Lumpable Partition

Proposition 5.1 generalizes to networks with arbitrary automorphisms which we illustrate at the example of the ring graph. When the model on the ring with nearest

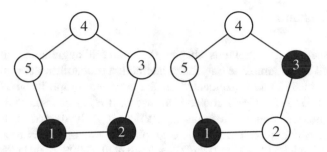

Fig. 5.19 Two configurations with equal $k = 2$ which are not macroscopically equivalent for the ring with $N = 5$

neighbor interactions is defined by $\omega(i, i+1) = \frac{1}{N} : i \bmod N$, it possesses an invariance with respect to translations. That is, the automorphism group $Aut_{\omega}(N)$ consists of all cyclic shifts of agents generated by $\sigma : (1, 2, \dots, N) \to (N, 1, 2, \dots, N-1)$. Notice that translational symmetries of this kind also play an important role in the determination of the relevant dimensions of spin rings (Bärwinkel et al. 2000) and that there are interesting parallels in between the two problems.

Consider a ring of five agents ($N = 5$) with $2^5 = 32$ micro states. For $\mathbf{x} = (\blacksquare\blacksquare\blacksquare\blacksquare\blacksquare)$ it is clear that $\sigma^k(\mathbf{x}) = \mathbf{x}$ for all k. That is, $\mathbf{x} = (\blacksquare\blacksquare\blacksquare\blacksquare\blacksquare)$ with $k = 0$ constitutes a class of its own. For $k = 1$, we may use $x_1 = (\square\blacksquare\blacksquare\blacksquare\blacksquare)$ as a generator (5.5) for its class. As all 5 configurations with $k = 1$ can be obtained shifting x_1, all of them are in the same equivalence class. The 10 configurations with $k = 2$ cannot be lumped into the same macro state. There are two classes differentiated by the distance of zero or one in between the two black agents (see Fig. 5.19). Using the two configurations shown in Fig. 5.19 as generators yields two equivalence classes each containing five micro states. The cases $k = 3, 4, 5$ follow by symmetry so that all in all the dimension $\tilde{\mathbf{X}}$ of macro chain is reduced to 8.

In the general case of N agents we can in principle proceed in the same way. However, the number of macro states will increase considerably with the system size. We finish this section with a quantification of this number for the ring for which we can use a well-known enumeration theorem due to Pólya (see Harary and Palmer 1973, pp. 35–45, Eqs. (2.2.10) and (2.4.15) in particular). According to this, the number of macro states is

$$|\tilde{\mathbf{X}}| = \frac{1}{N} \sum_{k|N} \varphi(k) 2^{\frac{N}{k}} \tag{5.13}$$

where $\varphi(k)$ is the Euler φ-function and the sum is over the divisors $k|N$ of N. As an approximation we have $|\tilde{\mathbf{X}}| \approx 2^N/N$. Hence, an explicit solution of the macro chain will be possible only for very small systems.

5.7　Discussion

We have seen in the previous chapter that the full aggregation illustrated in
Figs. 4.1 and 5.1 is lumpable only if the interaction probabilities are uniform. This
corresponds to the VM implemented on the complete graph in which $\omega(i,j) =$
$1/N(N - 1)$ (or $1/N^2$ if self-choice is allowed). It is, of course, well-known that
the macro model obtained in terms of $h(\mathbf{x}) = k$ fully describes the evolution
of the micro model on the complete graph, but not on other topologies, see
Slanina and Lavicka (2003, p. 3) and Castellano et al. (2009, p. 601). Nevertheless,
Proposition 5.1 sheds light on the (probabilistic) reason for this. Namely, the
complete graph and respectively homogeneous mixing is the only topology for
which the automorphism group is the group $\mathscr{S}_N$ of all permutations of N agents.
In this case, for any two configurations $\mathbf{x}, \mathbf{x}'$ with an equivalent aggregate k there
is a $\sigma \in \mathscr{S}_N$ such that $\mathbf{x} = \sigma(\mathbf{x}')$. Hence, an equivalent aggregate value k implies
macroscopic equivalence. The fact that this is only true for complete graph and
homogeneous mixing underlines how restrictive these conditions are.

　　The more complex the internal structure of the agents and the more heterogeneous their interaction behavior, the lower our chances to derive a loss-less coarse-
graining that leads to a tractable Markov chain. It is clear that in heterogeneous
networks with a small number of automorphisms the coarse-graining is limited
because only a few micro states are macroscopically equivalent and can be lumped.
As this method is based on exact graph automorphisms it is more suited for stylized
situations as the two-community and the leadership model discussed in Sects. 5.4
and 5.5.

　　On the other hand, the method informs us in this way about the complexity of a
system introduced by non-trivial interaction relations. Even in a model as simple as
the VM, the behavior of whole system is not completely described by summation
over its elements (full aggregation in terms of k), because non-trivial dynamical and
spatial effects may emerge at the macro level. In this sense, our work is related
to key concepts in the area of computational emergence (Bedau 2003; Huneman
and Humphreys 2008) dealing with criteria and proper definitions of emergence.
Thereafter "an emergent phenomenon is one that arises from a computationally
incompressible process" (Huneman and Humphreys 2008, pp. 425–426). Markov
projections as discussed here in the context of the VM provide explicit knowledge
about the (in)compressibility of computational models and may therefore help to
operationalize these rather abstract definitions. This issue is further discussed in
Chaps. 7 and 9, see also Banisch (2014).

　　Let us finally note that in general there may be many partitions $\mathscr{M}$ of the state
space that are lumpable and here no statement is made here about optimality of the
partition $\mathscr{M}_\omega$ generated by the application of Proposition 5.1. On the other hand,
a simple answer is provided by a closer inspection of the VM with homogeneous
mixing telling us that $\mathscr{M}_\omega$ is not optimal in that case. Namely, we have for any k,
$P(X_k, X_{k+1}) = P(X_{(N-k)}, X_{(N-k)\mp1})$ which means that the pairs $\{X_k, X_{(N-k)}\}$ can
be lumped into the same state. The reason for this is that the VM update rule

brings about an additional symmetry that is not accounted for in Aut_ω and therefore not in $\mathcal{M}_\omega$. More generally, the micro structure of the VM is always symmetric with respect to the simultaneous flip of all agent states $x_i \rightarrow \bar{x}_i, \forall i$ and therefore, independent of the interaction topology, $\hat{P}(\mathbf{x}, \mathbf{y}) = \hat{P}(\bar{\mathbf{x}}, \bar{\mathbf{y}})$.

References

Banisch, S. (2014). From microscopic heterogeneity to macroscopic complexity in the contrarian voter model. *Advances in Complex Systems, 17,* 1450025.

Banisch, S., & Lima, R. (2013). Markov chain aggregation for simple agent-based models on symmetric networks: The voter model. In *CoRR*. arxiv.org/abs/1209.3902 (to appear in Advances in Complex Systems).

Bärwinkel, K., Schmidt, H.-J., & Schnack, J. (2000). Structure and relevant dimension of the Heisenberg model and applications to spin rings. *Journal of Magnetism and Magnetic Materials, 212,* 240–250.

Bedau, M. A. (2003). Downward causation and the autonomy of weak emergence. *Principia Revista Internacional de Epistemologica, 6*(1), 5–50.

Castellano, C., Fortunato, S., & Loreto, V. (2009). Statistical physics of social dynamics. *Reviews of Modern Physics, 81*(2), 591–646.

Darroch, J. N., & Seneta, E. (1965). On quasi-stationary distributions in absorbing discrete-time finite Markov chains. *Journal of Applied Probability, 2*(1), 88–100.

Harary, F., & Palmer, E. M. (1973). *Graphical enumeration* (Vol. 16). New York: Academic.

Huneman, P., & Humphreys, P. (2008). Dynamical emergence and computation: An introduction. *Minds and Machines, 18*(4), 425–430.

Kemeny, J. G., & Snell, J. L. (1976). *Finite Markov chains*. Berlin: Springer.

Lazarsfeld, P., & Merton, R. K. (1954). Friendship as a social process: A substantive and methodological analysis. In M. Berger, T. Abel, & C. H. Page (Eds.), *Freedom and control in modern society* (pp. 18–66). New York: Van Nostrand.

Slanina, F., & Lavicka, H. (2003). Analytical results for the Sznajd model of opinion formation. *The European Physical Journal B - Condensed Matter and Complex Systems, 35*(2), 279–288.

Sood, V., & Redner, S. (2005). Voter model on heterogeneous graphs. *Physical Review Letters, 94*(17), 178701.

Wright, S. (1943). Isolation by distance. *Genetics, 28,* 114–138.

Chapter 6
Application to the Contrarian Voter Model

In this chapter, the results from the previous chapters are applied to the contrarian voter model (CVM from now on). This model is an extension of the VM which accounts for so-called contrarian behavior. Contrarian behavior relates to the presence of individuals that do not seek conformity under all circumstances or to the existence of certain situations in which agents would not desire to adopt the behavior or attitude of their interaction partner. In our case this shall be included into the model by introducing a small probability p with which agents to not imitate their interaction partner, but adopt precisely the opposite opinion. This leads to a non-absorbing random walk as the consensus profiles ($\square\square\cdots\square$) and ($\blacksquare\blacksquare\cdots\blacksquare$) are left with probability p.

In the next section, we shall briefly review the field of non-conformity opinion dynamics to which the CVM belongs. As with the VM in Chap. 3, we then start the analysis with a derivation of the micro-level description including a random mapping representation (RMR) associated to the CVM. Markov chain aggregation is then used to derive a macro chain for the complete graph as well as a meso-level description for the two-community graph composed of two (weakly) coupled sub-communities. The model dynamics are studies in terms of the contrarian rate p and the coupling r between the two communities with a particular focus on the stationary dynamics of the model. In both cases, a detailed understanding of the model behavior is possible using Markov chain tools.

6.1 Non-conformity Opinion Models

In most binary models of opinion dynamics (see Galam 2002; Sznajd-Weron 2004 for two well-known variants and Castellano et al. 2009 for an overview) a mechanism of local alignment leads to a system which converges to a final profile of global conformity (consensus) in which all agent share the same opinion. As an attempt to

© Springer International Publishing Switzerland 2016
S. Banisch, *Markov Chain Aggregation for Agent-Based Models*,
Understanding Complex Systems, DOI 10.1007/978-3-319-24877-6_6

make these models more realistic and avoid convergence to a fixed consensus profile in which no further opinion change is possible, several mechanisms to include nonconformity behavior into the models have been proposed. This includes agents that act independently of their interaction partner, inflexible agents that never change their mind and anti-conformity or contrarian behavior by which agents choose the opposite of their interlocutor.

The approach adopted for the CVM studied here is probably the most simple mechanism. In our choice, we basically follow Galam (2004) which, based on the concept of contrarian investment strategies in finance (Dreman 1980; Corcos et al. 2002), is the first study to introduce contrarian behavior into a model of opinion dynamics (namely, into Galam's majority model). While the majority model without contrarians is characterized by a relatively fast convergence to complete consensus, the introduction of only a small rate of contrarian choices leads to the coexistence of the two opinions with a clear majority-minority splitting. Noteworthy, as the contrarian rate increases further, the model exhibits a phase transition to a disordered phase in which no opinion dominates in the population. Similar observations have been made for the Sznajd model (de la Lama et al. 2005; Sznajd-Weron et al. 2011; Nyczka et al. 2012).

More recently the literature often distinguishes between two types of nonconformity: (1) anti-conformity or contrarian behavior and (2) independent or inflexible agents (Sznajd-Weron et al. 2011; Nyczka et al. 2012; Crokidakis et al. 2014). See Mobilia (2003), Galam and Jacobs (2007), Sznajd-Weron et al. (2011), Nyczka et al. (2012), Crokidakis et al. (2014), and Maity and Mukherjee (2016) for opinion models that include independent or inflexible agents. The importance of a distinction between individuals that generally oppose the group norm or act independently of it is, from the socio-psychological perspective, relatively obvious. The fact that these two behaviors may also give rise to qualitatively different dynamical properties, however, has been established only recently (Nyczka et al. 2012). Here we stick to contrarians.

The voter model with contrarians presented in Masuda (2013) is probably the one that relates most to the model used here. The main difference is that a fixed number of agents always acts in a contrarian way whereas in the present model all agents take contrarian choices with a small probability p. In that setting, Masuda (2013) could not observe the phase transition from majority-minority splitting to disorder, but rather a change from a uniform to a Gaussian equilibrium distribution. While this difference in comparison with the Sznajd and Galam models has been attributed to the linearity of the CVM in Masuda (2013), this chapter shows that there is in fact an order-disorder phase transition in the CVM as well. However, the ordered phase can be observed only below a very small contrarian rate of $p^* = 1/(N+1)$ at which the equilibrium distribution is uniform (see Sect. 6.3.2), in accordance with Masuda (2013). In the setting of Masuda (2013) with a fixed number of contrarian agents, however, this value is already reached, on average, with only a single contrarian, independent of the population size.

Notice, finally, that the complete graph plays an exceptional role for the analytical treatment of nonconformity models commended on above. From the

Markovian point of view, this is due to the fact that for the complete graph binary opinion models are lumpable—that is reducible without loss of information—to a macroscopic description in terms of the average opinion or "magnetizationy (see Chap. 3). This paper analyses the complete graph as well, but it goes beyond it by studying the CVM on a perfect two-community graph. According to the method introduced in Chap. 5, a loss-less macro description is obtained by taking into account separately the average opinion in the two sub-graphs, that is, by a refinement of the level of observation.

6.2 The CVM Micro Chain

6.2.1 Model

As the VM, which has been discussed at length in the previous chapters, the CVM is a binary opinion model where N agents can adopt two different opinions: □ and ■. The model is an extension of the VM in order to include a form of contrarian behavior. At each step, an agent (i) is chosen at random along with one of its neighbors (j). Usually (with probability $1 - p$), i imitates j, but there is also a small probability p that agent i will do the opposite (contrarian extension). More specifically, if i holds opinion ■ and meets an agent j in □, i will change to □ with probability $1 - p$, and will maintain its current state ■ with probability p. Likewise, if i and j are in the same state, i will flip to the opposite state with probability p.

While the VM rule may be interpreted as a kind of ferromagnetic coupling by which neighboring spins (agents) align, the contrarian rule can be interpreted as anti-ferromagnetic coupling by which neighbors are of opposed sign after the interaction. Table 6.1 illustrates the update rules for the CVM.

6.2.2 Micro Dynamics

From the micro-level perspective (see Sect. 3.2), the CVM implements an update function of the form $\mathbf{u} : \mathbf{S} \times \mathbf{S} \times \Lambda \to \mathbf{S}$. That is, the new state of a randomly chosen

Table 6.1 Update rules $y_i = \mathbf{u}(x_i, x_j)$ for the CVM

Prob.	x_j	x_j	$(1-p)$	■	□	p	■	□
x_i	y_i	y_i	■	■	□	■	□	■
x_i	y_i	y_i	□	■	□	□	□	■

The VM rule (ferromagnetic coupling) is applied with probability $(1 - p)$, the contrarian rule (anti-ferromagnetic coupling) with probability p

agent i is given by

$$y_i = \mathbf{u}(x_i, x_j, \lambda) = \left\{ \begin{matrix} x_j : \lambda = \lambda_V \\ \bar{x}_j : \lambda = \lambda_C \end{matrix} \right\}, \tag{6.1}$$

where $\bar{x}_j$ denotes the opposite attribute of x_j. In each iteration, two agents i, j are chosen along with a random variable $\lambda \in \Lambda = \{\lambda_V, \lambda_C\}$ that decides whether the voter (λ_V) or the contrarian rule (λ_C) is performed. The probability for that is $\omega(i, j, \lambda)$. Notice that the update rule is equal for all agents and independent from the agent choice. Therefore the probability that an agent pair (i, j) is chosen to perform the contrarian rule can be written as $\omega(i, j) Pr(\lambda = \lambda_C) = p\omega(i, j)$. Respectively, we have $(1 - p)\omega(i, j)$ for the VM rule.

Let us briefly discuss the random mapping representation of the CVM for three agents and complete connections (i.e., $\omega(i, j) = 1/N(N - 1)$, $\forall i \neq j$). As shown in Table 6.2, this model variant implements 12 mappings F_z, because for any of the 6 possible agent choices (i, j) there is an additional binary choice between the voter (λ_V) and the contrarian (λ_C) update mechanism. To index the set $\mathscr{F}_Z$ of mappings we now use a triple (i, j, λ) where i and j correspond to the random agent choice and $\lambda = \{\lambda_V, \lambda_C\}$ to the random choice of one or other rule. From Table 6.2 it is easy to see that the respective probability distribution $\omega(i, j, \lambda)$ is independent and identically distributed as it remains the same for any iteration at any time. That is to say, the micro-level process is a Markov chain.

The CVM with sequential update belongs to the class of single-step models for which the microscopic transition probabilities have been discussed in Sect. 3.2.3. Considering that λ_C (for contrarian rule) is chosen with probability p and λ_V (VM rule) with $(1 - p)$, and that this choice is independent of the agent choice, the CVM micro-level transition probability $\hat{P}(\mathbf{x}, \mathbf{y})$ between two adjacent configurations $\mathbf{x} \overset{i}{\sim} \mathbf{y}$

Table 6.2 $\mathscr{F}_Z$ for the CVM with three agents

z	(i, j, λ)	a ■■■	b ■■□	c ■□■	d □■■	e ■□□	f □■□	g □□■	h □□□
1	$(1, 2, \lambda_V)$	a	b	g	a	h	b	g	h
2	$(1, 2, \lambda_C)$	d	f	c	d	e	f	c	e
3	$(1, 3, \lambda_V)$	a	f	c	a	h	f	c	h
4	$(1, 3, \lambda_C)$	d	b	g	d	e	b	g	e
5	$(2, 1, \lambda_V)$	a	b	a	g	b	h	g	h
6	$(2, 1, \lambda_C)$	c	e	c	d	e	f	d	f
7	$(3, 1, \lambda_V)$	a	a	c	f	c	f	h	h
8	$(3, 1, \lambda_C)$	b	b	e	d	c	d	g	g
9	$(2, 3, \lambda_V)$	a	e	a	d	e	h	d	h
10	$(2, 3, \lambda_C)$	c	b	c	g	b	f	g	f
11	$(3, 2, \lambda_V)$	a	a	e	d	e	d	h	h
12	$(3, 2, \lambda_C)$	b	b	c	f	c	f	g	g

is given by

$$\hat{P}(\mathbf{x}, \mathbf{y}) = (1 - p) \sum_{j:(y_i = x_j)} \omega(i,j) + p \sum_{j:(y_i = \bar{x}_j)} \omega(i,j). \tag{6.2}$$

It is clear that, as for the original VM, the micro-level process for the CVM corresponds to a random walk on the hypercube. However, it is noteworthy that the CVM leads to a *regular* chain (as opposed to an absorbing random walk for the original VM). Namely, whenever $p > 0$, there is a non-zero probability that the process leaves the consensus states ($\square\square \ldots \square$) and ($\blacksquare\blacksquare \ldots \blacksquare$). Equation (6.2) tells us that this probability is precisely p. Therefore, the system does not converge to a fixed configuration and the long-term behavior of the model can be characterized by its stationary distribution.

6.3 Homogeneous Mixing

This section analyses the behavior of the CVM for homogeneous mixing. As seen in the previous chapters, the case of homogeneous mixing is particularly simple because the micro chain is lumpable with respect to the partition $\mathbf{X} = \{X_0, \ldots, X_k, \ldots, X_N\}$ (with $0 \le k \le N$) induced by the macroscopic measure that counts the number of agents in the different states. Therefore, important entities of interest (e.g., stationary distribution and mean passage times) can be computed on the basis of the respective transition matrix.

6.3.1 Macro Chain

Let us consider that the model is implemented on the complete graph without loops. In that case, the probability to choose a pair (i,j) of agents becomes $\omega(i,j) = 1/N(N-1)$ whenever $i \ne j$ and $\omega(i,i) = 0, \forall i$. As before, it is clear that the interaction structure is invariant with respect to all agent permutations (that is, $\omega(i,j) = \omega(\sigma i, \sigma j), \forall \sigma \in \mathscr{S}_N$ and all pairs (i,j)) and therefore all agent configurations with the same number k of agents in $\square$ (and therefore $N - k$ in $\blacksquare$) belong to the same class of macroscopic equivalence and can be mapped into the same macro atom (X_k). See Sect. 4.1.1 and Proposition 5.1. In other words, for homogeneous mixing *full aggregation* over all agents does not destroy Markovianity, which is in complete analogy to the pure VM. Notice again that in hypercube terminology that level of observation corresponds to the Hamming weight of a configuration $h(\mathbf{x}) = k$.

Consequently, since all agents interact with all the others with equal probability, the respective transition rates depend only on the numbers k and $N - k$ of agents in the two states. Consider, for example, the probability $P(X_k, X_{k+1})$ that a black agent flips its state. There are two situations in which this change can happen: first, if a pair of states $(x_i, x_j) = (\blacksquare, \square)$ along with VM update is chosen, i.e., $(x_i, x_j, \lambda) = (\blacksquare, \square, \lambda_V)$, second, if a pair $(\blacksquare, \blacksquare)$ is chosen along with contrarian update, i.e., $(x_i, x_j, \lambda) = (\blacksquare, \blacksquare, \lambda_C)$. In a configuration with k agents in $\square$, there are $(N - k)k$ possibilities for the first option and $(N - k)(N - k - 1)$ possibilities for the latter.[1]

Alternatively, $P(X_k, X_{k+1})$ can be obtained by evaluating the transition probability (6.2) from some $\mathbf{x} \in X_k$ to the set of $\mathbf{y} \in X_{k+1}$, denoted in the previous chapter as $\hat{p}_{\mathbf{x}, X_{k+1}}$ Then we obtain

$$
\begin{aligned}
P(X_k, X_{k+1}) &= \sum_{x_i = \blacksquare} \left[(1 - p) \sum_{j : (x_j = \square)} \omega(i, j) + p \sum_{j : (x_j = \blacksquare)} \omega(i, j) \right] \\
&= (N - k) \left[(1 - p)k\omega + p(N - k)\omega \right] \\
&= (1 - p)\frac{(N-k)k}{N(N-1)} + p\frac{(N-k)(N-k-1)}{N(N-1)}.
\end{aligned}
\tag{6.3}
$$

Similarly, we obtain for $P(X_k, X_{k-1})$

$$
P(X_k, X_{k-1}) = (1 - p)\frac{(N-k)k}{N(N-1)} + p\frac{k(k-1)}{N(N-1)}.
\tag{6.4}
$$

And finally,

$$
P(X_k, X_k) = \frac{k^2(2-4p) + 2kN(2p-1) + N(N-Np+p-1)}{N(N-1)}
\tag{6.5}
$$

Figure 6.1 aims at giving an intuition about the dynamical structure of the process by considering the relation between the probability for a transition one step to the right, $P(X_k, X_{k+1})$, and a transition to the left, $P(X_k, X_{k-1})$, as a function of k. This informs us about the more probable tendency for future evolution for every atom in the macro chain. Figure 6.1 shows that $P(X_k, X_{k+1}) > P(X_k, X_{k-1})$ for $k < N/2$ and respectively $P(X_k, X_{k+1}) < P(X_k, X_{k-1})$ for $k > N/2$ telling us that the contrarian rule (performed with probability p) introduces in every atom X_k a small bias that drives the system towards the fifty-fifty configurations. This bias is given by

$$
P(X_k, X_{k+1}) - P(X_k, X_{k-1}) = p - \frac{2kp}{N}.
\tag{6.6}
$$

---

[1]Notice that, contrary to the treatment of the VM in Chap. 3, we do not allow that agents interact with themselves. In the case self-choice is allowed ($\omega(i, i) > 0$), the number of possibilities for $(\blacksquare, \blacksquare)$ modifies to $(N - k)^2$. We also have $\omega(i, j) = 1/N(N - 1)$, $\forall i \neq j$ compared to $1/N^2$ in the model with self-choice. Even for systems of moderate size the dynamical effect of this slight difference in transition rates is neglectable.

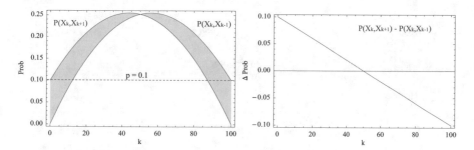

Fig. 6.1 Transition probabilities and difference in transition probabilities as function of k ($N = 100$)

6.3.2 Stationary Dynamics for Homogeneous Mixing

As already mentioned, contrary to the pure VM, the contrarian variant does no longer lead to an absorbing Markov chain, but results in a regular chain. In the case that the population reaches consensus ($k = 0$ or $k = N$) there is still a small probability, (namely $P(X_0, X_1) = P(X_N, X_{N-1}) = p$) with which the consensus configuration is left. In that case, a statistical understanding of the model behavior is provided by the limiting vector or stationary distribution it converges to. That is, by the distribution π that remains unchanged under further application of P:

$$\pi P = \pi. \tag{6.7}$$

Notice that Eq. (6.7) tells us that the stationary distribution π of a Markov chain $(\mathbf{X}, P)$ is proportional to the left eigenvector of P associated to the maximal eigenvalue $\lambda_{max} = 1$. It is well-known (Kemeny and Snell 1976, p. 69ff) that regular chains have a unique limiting vector and that the process converges to it for any initial distribution. Notice also, that the rate of convergence is usually related to the second largest eigenvalue of P ($\lambda_2 < 1$) in the sense that the order of convergence is proportional to λ_2^t (Kemeny and Snell 1976; Behrends 2000, among others).

For a system of 100 agents the stationary vector π is shown in Fig. 6.2 for various contrarian rates p. The horizontal axis represents the macro states X_k for $k = 0, 1, \ldots, N$ and the π_k correspond to the probability with which the process is in atom X_k provided it is run long enough and has reached stationarity. Notice that, for a large number of steps, the π_k also represent the expected value for the fraction of time the process is in X_k (Kemeny and Snell 1976, Sect. 4.2). On the bottom of Fig. 6.2, three characteristic time series (three single simulation runs) are shown, one for large, one for intermediate and one for low p values. This provides a better understanding of the meaning of the stationary vector in relation to the time evolution of the respective processes.

Two different regimes can be observed in Fig. 6.2 characterized by the green and red curves respectively. A large contrarian rate p (green curves) leads to a

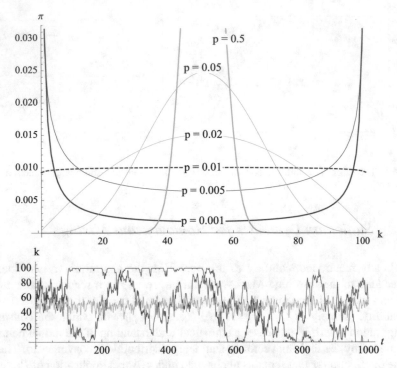

Fig. 6.2 Stationary vector of the CVM with $N = 100$ and homogeneous mixing for various p. There is a transition from the absorbing VM to the random fluctuations around the mean. On the *bottom*, the respective example time series are shown

process which fluctuates around the states with approximately the same number of black and white agents—the fifty-fifty situation ($k = N/2$) being the most probable observation. The larger p, the lower the probability to deviate strongly from the fifty-fifty configurations. In fact, the process resembles a random process in which agent states are flipped at random.

A different behavior is observed if p is small. This is represented by the red curves. For a small contrarian rate, the population is almost uniform (consensus) for long periods of time, but due to the random shocks introduced by the contrarian rule there are rare transitions between the two extremes. This is very similar to the VM at low (but non-zero) temperature, where random state switches or excitations take the role of mutations and prevent the system from complete freezing to the zero-temperature ground state. In between these two regimes, there is a $p \approx 0.01$ for which the process wanders through the entire state space, in such a way that the stationary distribution is almost uniform.

Figure 6.3 shows the situation for a system of 1000 agents. The same two regimes are observed for the larger system. However, the value p at which the behavior changes from the switching between two consensus states to fluctuations around the fifty-fifty situation is decreased compared to the $N = 100$ case. In the case of

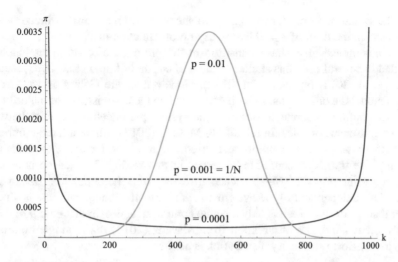

Fig. 6.3 Stationary vector of the CVM with 1000 agents. Notice the (almost) uniform stationary distribution when $p = 1/N$

$N = 1000$, an almost uniform stationary distribution is observed for $p \approx 0.001 = 1/N$. To be more precise, it is, in fact, not difficult to show that for any system size N the stationary distribution is uniform with $\pi_k = 1/(N+1)$, $\forall k$ exactly for $p^* = 1/(N+1)$. All that is necessary in order to verify this is to show that $\pi P = \pi$ in this case. Hence, we have to show that

$$\frac{1}{N+1} \left(P(X_{k-1}, X_k) + P(X_k, X_k) + P(X_{k+1}, X_k) \right) = \frac{1}{N+1} \tag{6.8}$$

which is satisfied whenever

$$P(X_{k-1}, X_k) + P(X_k, X_k) + P(X_{k+1}, X_k) = 1. \tag{6.9}$$

Notice that Eq. (6.9) is equivalent to requiring that P is a doubly stochastic matrix, and it is well-known that any doubly stochastic matrix has a uniform stationary vector. It is easy to show that for the CVM, Eqs. (6.8) and (6.9) are satisfied precisely for $p^* = 1/(N+1)$, but not for other contrarian rates.

When the contrarian rate p crosses the critical value $p^* = 1/(N+1)$, the system undergoes a continuous phase transition from majority-minority switching (ordered phase) to a balanced fifty-fifty situation in which no stable majorities form (disordered phase). The fact that $p^* = 1/(N+1)$ leads to $\pi_k = 1/(N+1)$, $\forall k$ shows the existence of large fluctuations at the critical contrarian rate, because the only way to have a stationary uniform distribution is to have very large fluctuations at any value of the state space. For large p, the system behaves around the mean value (here 50 and respectively 500) with only small deviations. For small p closed to 0, the system is rarely far from the two states of complete order (the consensus states) and in the limit of $p = 0$ has no asymptotic fluctuations at all.

The emergence of a phase transition in the presence of contrarians has been reported in several previous studies (Galam 2004; de la Lama et al. 2005; Sznajd-Weron et al. 2011; Nyczka et al. 2012). However, in the Galam as well as the Sznajd model the ordered phase is characterized by the co-existence of majority and minority opinions whereas in our case the system permanently switches between the two consensus profiles. In relation to Masuda (2013), where a fixed number of agents that always act as contrarians is used, the result derived here explains why such a phase transition cannot be observed in this setting. The reason is that the transition value $p^* = 1/(N + 1)$ scales inversely with the population size in such a way that, independent of the system size, the effective contrarian rate is already larger than p^* even if there is only a single contrarian agent (in that case $p = 1/N$). On the other hand, the observation from Masuda (2013) that a single contrarian leads to a uniform stationary distribution is confirmed.

6.3.3 Rate of Majority-Minority Switching

One of the most interesting advantages of using Markov chains as a macro description of the model is that it facilitates the computation of a series of quantities that one might wish to look at and which are more difficult to assess with other techniques. For the CVM, for instance, we can look at the mean number of steps required to go from one consensus state to the opposite consensus state. As in the absorbing case, the key to this (and to several other) computations is a matrix called the fundamental matrix (Kemeny and Snell 1976, p. 75ff). For regular chains it is computed by

$$\mathbf{F} = (\mathbf{I} - (P - W))^{-1} \tag{6.10}$$

where W is the limiting matrix with all rows equal to π (note that, $\lim_{n \to \infty} P^n = W$). Following Kemeny and Snell (1976), the fundamental matrix can be used to compute another matrix M which contains the mean number of steps between two states, say i and j, for any pair of states:

$$M = (\mathbf{I} - \mathbf{F} + \mathbf{EF}_{diag})D \tag{6.11}$$

where $\mathbf{E}$ is a matrix with all elements equal to one, $\mathbf{F}_{diag}$ the diagonal fundamental matrix with $(\mathbf{F}_{diag})_{ii} = (\mathbf{F})_{ii}$; $(\mathbf{F}_{diag})_{ij} = 0$, and D the diagonal matrix with $(D)_{ii} = 1/\pi_i$. The mean time from one consensus state to the other is then given by the element $M(0, N)$ which is plotted in Fig. 6.4 for system sizes from $N = 100$ to $N = 500$.

Fig. 6.4 Mean number of steps required to go from one to the other consensus state as a function of the scaled contrarian rate $(N + 1)p$

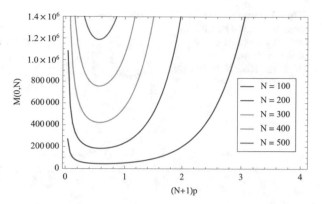

Notice that in Fig. 6.4 the contrarian rate p is scaled by the size of the macro chain $(N + 1)$ in order to compare the different cases. This accounts for the above-mentioned fact that the "critical" parameter value p^* at which a uniform stationary distribution is found depends inversely on the number of agents as $p^* = 1/(N + 1)$. Consequently, in Fig. 6.4, the uniform case is represented by $(N + 1)p = 1$. The switching behavior (from one consensus to the other and back) is found for values below that and the behavior approaches the random regime for values larger than one.

We observe in Fig. 6.4 that transitions between the two different consensus states are most frequent for a contrarian rate that is slightly below the "critical" contrarian rate p^*. There is a trade-off between the probability to indeed enter the state of complete consensus and the probability to go away from that and approach to the other extreme. For the p-values where $M(0, N)$ is minimal, both probabilities are relatively high. As contrarian rate decreases, the probability to reach consensus increases significantly, but a transition to the opposite consensus state is becoming rare. On the other hand, when p increases slightly, transitions from $k \approx 0$ to $k \approx N$ and back are still rather likely, but in many case the process turns in direction before a complete ordering has been achieved. This is true also for $p \approx p^*$. As p increases further, there is a strong decrease in probability to reach consensus altogether (see Fig. 6.2) and therefore the mean time between the two consensus states increases tremendously.

Finally, Fig. 6.5 shows the same analysis for transitions between states with a strong majority of □-agents to an equally strong majority of ■-agents. The same qualitative behavior is observed in the sense that switching between strong majorities ($X_0 \leftrightarrow X_N, X_5 \leftrightarrow X_{95}, X_{10} \leftrightarrow X_{90}$) becomes rather unlikely as the contrarian rate increases. On the other hand, transitions between moderate majorities of different sign (80 % and respectively 67 %) occur rather frequently and the contrarian rate at which the mean time between them becomes minimal is larger.

Fig. 6.5 Mean number of
steps required to go from X_k
to X_s as a function of the
scaled contrarian rate
$(N + 1)p$. Here $N = 100$

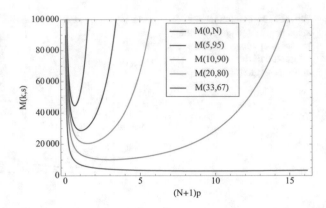

6.4   Two-Community Model

In this section, we consider the CVM on a two-community graph where the size of
the two communities is given by $M = L = 50$. The pure VM on two communities
has been discussed in the previous chapter (Sect. 5.4) and for the CVM the same
procedure can be used to obtain a Markov projection by strong lumpability; namely,
Proposition 5.1.

6.4.1   Meso Chain

In particular, the description of the interaction topology in Sect. 5.4.1 and derivation
of the respective interaction probabilities ω is in complete analogy to the previous
chapter. That is,

$$\gamma = \frac{r}{2LM + ((L-1)L + (M-1)M)r}$$

$$\alpha = \frac{1}{2LM + ((L-1)L + (M-1)M)r}, \tag{6.12}$$

where γ is the probability of intra-community interaction (strong ties) and α the
probability of inter-community interaction (weak ties), and $r = \alpha/\gamma$ the ratio
between the two.

As described in Sect. 5.4.2 for the pure VM, the interaction probabilities ω
defined by Eq. (6.12) give rise to a symmetry group $Aut_\omega(N) = (1\ldots M)(M +
1\ldots N)$ and Proposition 5.1 tells us that Markovianity is preserved by a projection
onto the $(M + 1) \times (L + 1)$ lattice. Each lattice point $\tilde{X}_{m,l}$ is associated to the
attribute frequencies m and l within the two sub-communities. In other words, the
model dynamics can be captured without loss of information by a "mesoscopic"

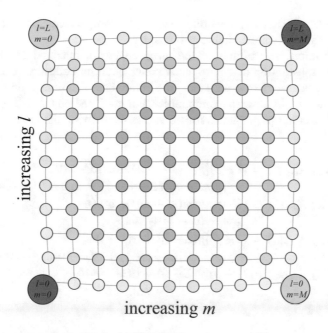

Fig. 6.6 The structure of the CVM meso chain for $L = M = 10$. The consensus states $\tilde{X}_{0,0}, \tilde{X}_{M,L}$ as well as the states of inter-community polarization $\tilde{X}_{0,L}, \tilde{X}_{M,0}$ are *highlighted*. The stationary distribution is mapped into node colors from *blue* (low values) to *red* (high values)

formulation in terms of attribute frequencies m and l in the two communities. The state space of the projected model is visualized in Fig. 6.6.

The colors shown in Fig. 6.6 represent the stationary distribution of the CVM with a relatively small contrarian probability p and a very weak coupling between the two islands. The large atoms in the corners of the grid highlight the states that represent configurations of high order. On the one hand (red-shaded in Fig. 6.6) there are the consensus configuration with all agents in the same state: $\tilde{X}_{L,M}$ and $\tilde{X}_{0,0}$. On the other hand (yellow-shaded), we have the states in which all agents of the same sup-group are aligned, but there is a disagreement across the sub-groups: $\tilde{X}_{0,M}$ and $\tilde{X}_{L,0}$. As before, we refer to these states as inter-community polarization.

In what follows, we shall refer to the chain shown in Fig. 6.6 (obtained via strong lumpability) as *meso chain* and denote the state space as $\tilde{X} = (\tilde{X}_{0,0}, \ldots, \tilde{X}_{m,l}, \ldots, \tilde{X}_{M,L})$. The notion of "meso" in this context accounts for the fact that the process $(\tilde{X}, \tilde{P})$ is indeed in between the micro and the macro level. Namely, it is a strong reduction compared to the microscopic chain $(\Sigma, \hat{P})$, but the number of states is still considerably larger than the macro system $(\mathbf{X}, P)$ obtained by aggregation over the entire agent population ($h(\mathbf{x}) = k$). While the full aggregation compatible with homogeneous mixing has lead to a random walk on the line with $N + 1 = O(N)$ states, the two-community model leads to a random walk on a 2D lattice with $O(N^2)$ states. Noteworthy, the latter is a proper refinement of the former, a fact that shall be exploited in the next chapter.

The transition probabilities of the meso chain are obtained on the basis of Eq. (6.2) by substitution of the respective interaction probabilities (6.12). That is, $\omega(i,j) = \gamma$ whenever two agent i and j are in the same community and $\omega(i,j) = \alpha$ whenever they are in different communities. For the CVM on two islands of size M and L the transition probabilities for the transitions leaving the atom $\tilde{X}_{m,l}$ are then given by

$$
\begin{aligned}
\tilde{P}(\tilde{X}_{m,l}, \tilde{X}_{m+1,l}) &= (1-p)[\gamma(m(M-m)) + \alpha(M-m)l] \\
&\quad + p[\gamma(M-m)(M-m-1) + \alpha(L-l)(M-m)] \\
\tilde{P}(\tilde{X}_{m,l}, \tilde{X}_{m-1,l}) &= (1-p)[\gamma(m(M-m)) + \alpha m(L-l)] \\
&\quad + p[\gamma m(m-1) + \alpha lm] \\
\tilde{P}(\tilde{X}_{m,l}, \tilde{X}_{m,l+1}) &= (1-p)[\gamma(L-l)l + \alpha(L-l)m] \\
&\quad + p[\gamma(L-l)(L-l-1) + \alpha(L-l)(M-m)] \\
\tilde{P}(\tilde{X}_{m,l}, \tilde{X}_{m,l-1}) &= (1-p)[\gamma(L-l)l + \alpha(M-m)l] \\
&\quad + p[\gamma l(l-1) + \alpha lm].
\end{aligned}
\tag{6.13}
$$

6.4.2   Stationary Dynamics on the Two-Community Graph

As described in Sect. 6.3.2, the stationary distribution π of a Markov chain with transition matrix P is the probability vector that satisfies $\pi P = \pi$ so that the computation of π requires the computation of the left eigenvector of P. The Markov projection of the two-community model with $M = L = 50$ results in a Markov chain of size $(M + 1)(L + 1) = 2601$. For a matrix of size 2601×2601 the (numerical) solution of the corresponding eigenvalue problem is still possible, but increasing the number of agents (that is, M and L) will soon lead to matrix sizes for which the solution for eigenvalues and vectors is rather costly.

There are two parameters that decide about the dynamical behavior of the CVM on the two-community graph: (1) the contrarian rate p, and (2) the coupling between the two islands captured by $r = \alpha/\gamma$. To obtain a complete picture of the model dynamics, the stationary distribution has been computed for various different values p and r which is shown in Fig. 6.7. From the top to the bottom, p is increased from $p = 0.01$, $p = 0.015$, $p = 0.02$ to $p = 0.03$. The plots in the left-hand column show the result for a moderate coupling between the two island with $r = 1/100$. A reduced coupling of $r = 1/1000$ is shown in the plots in the right-hand column.

The comparison of the left- and the right-hand side of Fig. 6.7 shows that the stationary probability for states of inter-community polarization, as well as the states close to them, increases with a decreasing coupling between the communities. That is, the configurations with intra-community consensus, but disagreement across the

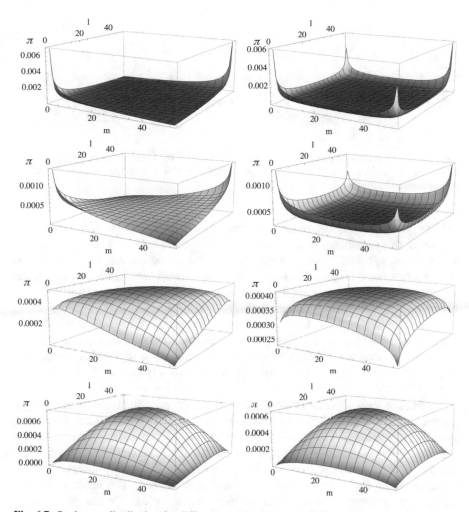

Fig. 6.7 Stationary distribution for different p and r for a system of $M = L = 50$. The column on the *l.h.s.* is for a moderate coupling $r = 1/100$ and the four plots on the *r.h.s.* are for a weak coupling $r = 1/1000$. From *top* to *bottom* the contrarian rates are $p = 0.01, 0.015, 0.02, 0.03$. The stationary probability for the consensus states ($m = l = 0$ and $m = l = 50$) increases with decreasing p. The stationary probability for the states of partial order ($m = M, l = 0$ and $m = 0, l = L$) increases as the coupling between the island r decreases. This topological effect is undermined by an increasing contrarian rate p

communities become more and more probable. This is very obvious for the plots with a small contrarian rate $p = 0.01$ and $p = 0.015$ where the probability to observe the states $\tilde{X}_{0,50}$ or $\tilde{X}_{50,0}$ becomes very high when decreasing the coupling to $r = 1/1000$. In fact, all configurations in which consensus is formed in at least one of the communities are rather likely (values along the border of the surface) whereas disordered configurations are rare. This is in direct analogy to the pure

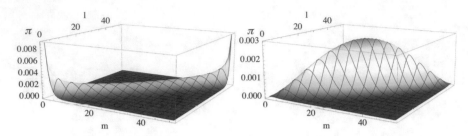

Fig. 6.8 Stationary distribution for $p = 0.01$ and $p = 0.03$ for a system of $M = L = 50$ and $\alpha = \gamma$ (homogeneous mixing). States of partial order ($m = M, l = 0$ and $m = 0, l = L$) become a very rare random event

VM ($p = 0$) and the corresponding quasi-stationary distribution (see Fig. 5.11). However, this significant difference between a moderate ($r = 1/100$) and a very weak ($r = 1/1000$) coupling diminishes as the contrarian rate becomes larger. This second trend observed in Fig. 6.7 is in agreement with what happens in the homogeneous mixing case as the contrarian rate p increases: the probability to observe consensus configurations with all agents in equal state becomes more and more unlikely and it is more and more likely to observe disordered agent configurations all together. In fact, a further increase of the contrarian rate will lead to a behavior that is essentially random and insensitive to topological constraints since the consensus formation within the communities is frequently perturbed by random events.

In order to show that a decreasing inter-community coupling leads generally to an increased stationary probability of intra-community polarization (local alignment, global polarization), let us compare the previous cases to the homogeneous mixing situation ($r = 1$). This is shown in Fig. 6.8 for the (relatively small) contrarian rates $p = 0.01$ and $p = 0.03$. It becomes clear that for $\alpha = \gamma$ states of partial order ($m = M, l = 0$ and $m = 0, l = L$) and states close to that become in effect exceptionally rare random events. The reason for this is clear. The number of micro configurations $\mathbf{x} \in \Sigma$ mapped into the state $\tilde{X}_{m,l}$ is $\binom{M}{m}\binom{L}{l}$ which is a huge number for $m \approx M/2, l \approx L/2$ but only 1 for $m = M, l = 0$ and $m = 0, l = L$. Because under homogeneous mixing there is no favoring of particular agent configurations with the same $k = m + l$ the stationary probability at meso scale is proportional to the cardinality of the set $\tilde{X}_{m,l}$.

6.5 Discussion

This chapter has provided an analysis of the CVM on the complete and the two-community graph. Based on the previous chapters, higher-level Markov chain descriptions have been derived and allow a detailed understanding of the two cases. A large contrarian rate p leads to a process which fluctuates around the states with approximately the same number of black and white agents, the fifty-fifty situation $k = N/2$ being the most probable observation. This is true for homogeneous mixing as well as for the two-community model. However, if p is small, a significant difference between the two topologies emerges as the coupling between the two communities becomes weaker. On the complete graph the population is almost uniform for long periods of time, but due to the random perturbations introduced by the contrarian rule there are rare transitions between the two consensus profiles. On the community graph, an effect of local alignment is observed in addition to that, because the system is likely to approach a meta-stable state of intra-community consensus but inter-community polarization.

A order-disorder phase transition as the contrarian rate increases has been observed on the complete graph in several previous contrarian opinion models (e.g., Galam 2004; de la Lama et al. 2005; Sznajd-Weron et al. 2011; Nyczka et al. 2012). For the CVM, in the transition from consensus switching to disorder there is a phase in which the process leads uniform stationary distribution in which all opinion frequency levels $0 \leq k \leq N$ are observed with equal probability ($\pi_k = 1/(N+1)$). The contrarian rate p at which this happens is $p^* = 1/(N+1)$ and depends inversely on the system size such that a model with a single contrarian agent fails to enter the ordered regime. This confirms and explains the behavior observed in Masuda (2013) for a model with a fixed number of contrarian agents.

It would be very interesting to perform the two-community analysis for other opinion models and different kinds of nonconformity behavior. For the Galam and the Sznajd model with contrarians (see Galam 2004; Nyczka et al. 2012) we hypothesize a more interesting distribution in which the four peaks are located at mixed minority-majority configurations. Furthermore, it would be interesting to analyze the effect of independent agents as studied in the q-voter model in Nyczka et al. (2012). Namely, on the complete graph (for $q > 5$) a third peak emerges centered at the fifty-fifty profiles and it is not clear how this effect translates on the two-community graph.

In the next chapter, we will stay with the CVM and focus on the effect of inhomogeneities in the interaction topology on the model behavior. We have seen in this chapter that homogeneous mixing compatible with the usual way of aggregation over all agents leads to a random walk on the line with $N+1 = O(N)$ states whereas the two-community model leads to a random walk on a 2D lattice with $O(N^2)$ states. As the latter is a proper refinement of the former this will give us means to study the relation between the two coarse-grainings in a Markov chain setting. The next chapter will show that the two-community CVM serves as a suitable scenario to assess the macroscopic effects introduced by a slight microscopic heterogeneity.

References

Behrends, E. (2000). *Introduction to Markov chains with special emphasis on rapid mixing*. Berlin: Friedrick Vieweg & Sohn.

Castellano, C., Fortunato, S., & Loreto, V. (2009). Statistical physics of social dynamics. *Reviews of Modern Physics, 81*(2), 591–646.

Corcos, A., Eckmann, J.-P., Malaspinas, A., Malevergne, Y., & Sornette, D. (2002). Imitation and contrarian behaviour: Hyperbolic bubbles, crashes and chaos. *Quantitative Finance, 2*(4), 264–281.

Crokidakis, N., Blanco, V. H., & Anteneodo, C. (2014). Impact of contrarians and intransigents in a kinetic model of opinion dynamics. *Physical Review E, 89*, 013310.

de la Lama, M. S., López, J. M., & Wio, H. S. (2005). Spontaneous emergence of contrarian-like behaviour in an opinion spreading model. *Europhysics Letters, 72*(5), 851.

Dreman, D. (1980). *Contrarian investment strategy: The psychology of stock market success*. New York: Random House.

Galam, S. (2002). Minority opinion spreading in random geometry. *The European Physical Journal B, 25*, 403–406.

Galam, S. (2004). Contrarian deterministic effects on opinion dynamics: "The hung elections scenario". *Physica A: Statistical Mechanics and its Applications, 333*(C), 453–460.

Galam, S., & Jacobs, F. (2007). The role of inflexible minorities in the breaking of democratic opinion dynamics. *Physica A: Statistical Mechanics and its Applications, 381*, 366–376.

Kemeny, J. G., & Snell, J. L. (1976). *Finite Markov chains*. Berlin: Springer.

Maity, S. K., & Mukherjee, A. (2016). Emergence of dominant opinions in presence of rigid individuals. In *Towards a theoretical framework for analyzing complex linguistic networks* (pp. 279–295). Berlin: Springer.

Masuda, N. (2013). Voter models with contrarian agents. *Physical Review E, 88*, 052803.

Mobilia, M. (2003). Does a single zealot affect an infinite group of voters? *Physical Review Letters, 91*(2), 028701.

Nyczka, P., Sznajd-Weron, K., & Cisło, J. (2012). Phase transitions in the q-voter model with two types of stochastic driving. *Physical Review E, 86*(1), 011105.

Sznajd-Weron, K. (2004). Dynamical model of Ising spins. *Physical Review E, 70*(3), 037104.

Sznajd-Weron, K., Tabiszewski, M., & Timpanaro, A. M. (2011). Phase transition in the Sznajd model with independence. *Europhysics Letters, 96*(4), 48002.

Chapter 7
Information-Theoretic Measures for the Non-Markovian Case

This chapter is devoted to the study of a non-Markovian case building upon the analysis of the contrarian voter model (CVM) discussed in the previous chapter. As noted earlier, two things may happen by projecting the microscopic Markov chain associated to an agent-based model (ABM) onto a coarser partition. First, the macro process is still a Markov chain which is the case of lumpability discussed most extensively throughout this book. Secondly, Markovianity may be lost after the projection induced by a certain observable which means that memory effects are introduced at the macroscopic level. This is a fingerprint of emergence in models of self-organizing systems. Noteworthy, in ABMs as well as more generally in Markov chains, this situation is the rule rather than an exception (Chazottes and Ugalde 2003; Gurvits and Ledoux 2005; Banisch et al. 2012; Banisch 2014).

In order to better understand the transition from the most informative "atomic" level to the levels at which the system behavior is typically observed we draw on the information-theoretic framework developed in Pfante et al. (2014a). We show that information-theoretic measures such as conditional past-future mutual information (Görnerup and Jacobi 2008) and micro-to-macro information flow (Pfante et al. 2014a) provide a reasonable framework to quantify the memory effects that are introduced by a global aggregation over the agent population without sensitivity to micro- or mesoscopic structures. The two-community CVM provides us with a suitable scenario to study how microscopic inhomogeneities in ABMs may lead to macroscopic complexity when the aggregation procedure defines a non-Markovian macro process. Using this scenario, we will first consider in detail the non-Markovianity of the model with respect to global aggregation and then quantify the emergent memory effects using the Markovianity measure (Görnerup and Jacobi 2008) and informational closure (Pfante et al. 2014a). This is a first step to gain a better insight into the principal microscopic conditions and mechanisms responsible for temporal correlation patterns observed at aggregate levels.

© Springer International Publishing Switzerland 2016
S. Banisch, *Markov Chain Aggregation for Agent-Based Models*,
Understanding Complex Systems, DOI 10.1007/978-3-319-24877-6_7

We shall begin this chapter with an introductory section which introduces different information-theoretic measures for multi-level systems and recalls the relation between those measures and (non)-lumpability. We then turn to the CVM example and describe the macroscopic process that is obtained by *full aggregation* over the attributes of the entire agent population. It is shown that different interaction networks may have a strong effect on the macroscopic stationary behavior of the CVM depending essentially on the propensity to foster local opinion clustering. In Sects. 7.3, 7.4 and 7.5 we focus on the two-community CVM. We first consider in detail why lumpability (in the strong as well as in the weak form) fails. After that we quantify the deviations from an idealized Markovian description using conditional past-future mutual information (intra-level deviation from Markovianity) and micro-to-macro information flow (inter-level flow) (mainly following Görnerup and Jacobi 2008 and Pfante et al. 2014a).

7.1 Lumpability and the Notion of Closure

Lumpability and information-theoretic closure measures provide two strongly interrelated perspectives to look at the relation between different levels of description in complex multi-level systems. Let us briefly reconsider these different notions here as an entry point into this chapter.

7.1.1 Information Measures for Multi-Level Systems

ABMs are an attempt to understand how macroscopic regularities emerge through processes of self-organization in systems of interacting agents. Aggregation and a multi-level perspective are key features in this modeling strategy. In addition to the prescribed individual level and the levels associated to the macroscopic observables, clustering and local alignment may lead to a hierarchy of intermediate levels which can be distinguished according to temporal and spatial scales and levels of aggregation. One of the main motivations for this book has been an improved theoretical understanding of the relation between different levels. In this chapter we study this relation for the case of non-Markovian aggregation by applying information-theoretic approaches for level identification in complex multi-level systems to ABMs.

The mathematical analysis of the relation between different levels of description is at the heart of statistical mechanics and has recently received some attention in the complex systems literature (Shalizi and Moore 2003; Görnerup and Jacobi 2010, 2008; Jacobi and Görnerup 2009; Pfante et al. 2014a,b). Consider a dynamical system $\hat{P} : \Sigma \to \Sigma$ with state space Σ and a Markovian (or deterministic) transition kernel $\hat{P}$. Further, consider an operator $\psi : \Sigma \to X$ that projects the dynamics onto a partition X of Σ inducing a dynamical process $(\hat{P})$ on X. As before, it is completely convenient to think of ϕ as a measurement process, a coarse-graining or aggregation of the original state space such that all states $\mathbf{x}, \mathbf{x}' \in \Sigma$ which give rise to the same measurement $\phi(\mathbf{x}) = \phi(\mathbf{x}')$ are mapped into the same state in $\mathbf{X}$. It is therefore also convenient to think of Σ and $\mathbf{X}$ as micro- and macroscopic state spaces respectively.

There are two interrelated ways of looking at such a multi-level description. The first one is related to *level identification* in cases where a higher level description is not given a priori, the second one with *level qualification* if a desired higher level of description is given beforehand by the level of observation that the problem at hand calls for, as often in AB research. Level identification tackles the problem of finding projection operators ϕ which lead to a "closed" description, in the sense that the system can be modeled by the state variables at the resulting $\mathbf{X}$-level without loss of information. In level qualification Σ, ϕ and consequently $\mathbf{X}$ are pre-defined and one may ask questions about the properties of the higher level process with respect to the original one. Notice that in many cases these two ways go hand in hand because the desired aggregate description is also the one which captures the dynamical details appropriately (see Shalizi and Moore 2003 for a discussion of this relation). Moreover, the task of level identification typically also involves the qualification of the different levels induced by the various different projection operators $\phi_\varepsilon : \Sigma \to \mathbf{X}$ because the induced higher-level descriptions are evaluated with respect to an idealized "closed" description (Pfante et al. 2014a,b).

A series of information-theoretic measures and concepts have recently proven to provide a suitable framework for the quantification of "closure" in multi-level systems (Görnerup and Jacobi 2008; Pfante et al. 2014a,b):

1. Markovianity: the induced higher-level dynamics P is Markovian. Deviation from Markovianity is quantified by the conditional mutual information $I(X_{t+1} : X_{-\infty}^{t-1}|X_t)$ (Shalizi and Moore 2003; Görnerup and Jacobi 2008; Pfante et al. 2014a).
2. Informational closure: a level is informational closed if the knowledge of the micro state $\mathbf{x}_t$ does not allow for better predictions of X_{t+1} than the knowledge of X_t, i.e. $I(X_{t+1}; \mathbf{x}_t|X_t) = 0$ (Pfante et al. 2014a,b).
3. Predictive efficiency: quantifies the predictive information of a process in relation to its complexity. In Shalizi (2001) the notion is introduced as the ratio between excess entropy and statistical complexity measured in terms of the entropy (Shalizi 2001; Pfante et al. 2014b).
4. Commutativity: meaning that there exists some transition kernel such that the diagram (Fig. 7.1) commutes (Pfante et al. 2014a).
5. Observational commutativity: it makes no difference, whether we perform the aggregation first, and then observe the upper process, or we observe the process on the micro state level, and then lumping together the states (Pfante et al. 2014a).

A comprehensive study of the relation between different measures has been presented in Pfante et al. (2014a). It has been shown, among many other things, that informational closure implies commutativity, observational commutativity and Markovianity. So far, the measures have been mainly applied to the evaluation of different coarse-grainings of simple dynamical systems with non-trivial dynamics

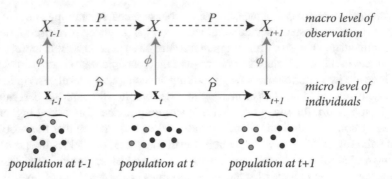

Fig. 7.1 A simple illustration of the multi-level perspective

(Görnerup and Jacobi 2008; Pfante et al. 2014b) with the focus to systematically explore different partitions. Görnerup and Jacobi (2008) apply their Markovianity measure to evaluate all possible partitions of a discretized version of the roof map as well as a 4-state automata and identify those partitions that lead to Markovian dynamics. In Pfante et al. (2014b), explicitly compute the Markovianity measure, informational closure and predictive efficiency for different coarse gainings of the tent map. In Pfante et al. (2014b) Markovianity, informational closure and predictive efficiency have been computed for different coarse gainings of the tent map.

For the purposes of this chapter—namely, the quantification of memory effects generated through a non-Markovian aggregation procedure—we will concentrate on the first and the second: Markovianity and informational closure.

The particular role of *Markovianity* in the definition or identification of macroscopic observables has been emphasized in Shalizi and Moore (2003). Based on that, Görnerup and Jacobi (2008) propose a Markovianity measure based on the idea is that an higher level is closed if the dynamic $P : \mathbf{X} \to \mathbf{X}$ induced at this level is Markovian. It is clear that this relates directly to the idea of lumpability. The aim of the Markovianity measure is to quantify how much information about the next symbol (X_{t+1}) is on average over all symbols contained in the sequence of symbols $(X_{-\infty}^{t-1} = [\ldots, X_{t-2}, X_{t-1}])$ before the current symbol (X_t), or, in other words, how much does knowledge of the past reduce uncertainty about X_{t+1} given X_t. In the Markovian case the conditional past future mutual information

$$I(X_{t+1} : X_{-\infty}^{t-1}|X_t) = H(X_{t+1}|X_t) - H(X_{t+1}|X_{-\infty}^{t-1}) \tag{7.1}$$

vanishes because looking further back into the past (denoted as $X_{-\infty}^{t-1}$) does not provide any new information about the future evolution. Noteworthy, the Markovianity measure can be expressed in terms of the slope of block entropies which bears an relation to process reconstruction in turbulence and finance (Chazottes et al. 1998; Vilela Mendes et al. 2002) which we will come back to in Sect. 7.4.

According to the measure of *informational closure*, introduced in Pfante et al. (2014a), a level is informational closed if the knowledge of the micro state $\mathbf{x}_t$ does not allow for better predictions of X_{t+1} than the knowledge of the previous macro state X_t. This is quantified by the information flow from the original to the higher level

$$I(X_{n+1} : \mathbf{x}_n | X_n) = H(X_{n+1} | X_n) - H(X_{n+1} | \mathbf{x}_n) \qquad (7.2)$$

which vanishes whenever the knowledge of the micro state x_n does not reduce uncertainty about the macroscopic evolution compared to the macro level prediction (i.e., when $H(X_{t+1} | X_t) = H(X_{t+1} | \mathbf{x}_t)$). As shown in Pfante et al. (2014a), $I(X_{t+1} : X_{-\infty}^{t-1} | X_t) \leq I(X_{t+1}; \mathbf{x}_t | X_t)$ so that vanishing information flow from micro to macro implies Markovianity.

In Sects. 7.4 and 7.5 , we will compute these two measures with respect to the global aggregation for the two-community CVM. This model provides a simple toy model of a multi-level system since, in addition to the micro and macro scale, an intermediate (lumpable) level exists (see Chaps. 5 and 6). But before, we will discuss the relation between the two closure measures and lumpability.

7.1.2   Lumpability and Closure

As before, we denote as $(\Sigma, \hat{P})$ a general micro chain with state space Σ and transition probabilities $\hat{P} : \Sigma \to \Sigma$ corresponding to an ABM and assume a projection operator $\phi : \Sigma \to \mathbf{X}$ that projects the chain onto a macroscopic level of interest which corresponds to a partition of Σ denoted as $\mathbf{X}$. Intuitively, as discussed in detail throughout the previous chapters, lumpability requires that the transition probability between pairs of macro states $X_k, X_l \in \mathbf{X}$ is independent of the particular microscopic configuration $\mathbf{x} \in X_k$. One way in which this may be satisfied is that for all pairs $X_k, X_l \in \mathbf{X}$ the probability to go to X_l is equal for all micro states $\mathbf{x}, \mathbf{y}$ within X_k. This corresponds to the case of strong lumpability.

From the point of view of closure, it is clear that in the case of lumpability the macro process $(\mathbf{X}, P)$ obtained by the lumpable projection provides a closed description of the original process in the sense that all information about the original dynamics is contained in the macroscopic description. For the Markovianity measure proposed by Görnerup and Jacobi (2008) (7.1) this is relatively obvious for it captures precisely the very idea of lumpability. According to this the higher level is closed if the dynamic $P : \mathbf{X} \to \mathbf{X}$ induced at this level is Markovian, which is satisfied by definition for any lumpable construction (for strong and for weak lumpability). More precisely, the macro-level conditional past future mutual information $I(X_{t+1} : X_{-\infty}^{t-1} | X_t)$ vanishes whenever $(\mathbf{X}, P)$ is again a Markov chain for the past $(X_{-\infty}^{n-1})$ does not provide any additional information about the future evolution. While lumpability is a rather restrictive "yes-no-question" either satisfied or not, Markovianity measured by $I(X_{t+1} : X_{-\infty}^{t-1}|X_t)$ allows to quantify to what extend a derived macro process deviates from Markovianity. It is therefore a measure of the memory effects introduced at the aggregate level by a specific aggregation map ϕ.

Vanishing information flow (7.2) for the case of strong lumpability is also relatively obvious. To see this let us look at the conditional entropies used in the computation of the measure:

$$H(X_{t+1}|X_t) = -\sum_{X_t \in \mathbf{X}} p(X_t) \sum_{X_{t+1} \in \mathbf{X}} p(X_{t+1}|X_t) \log p(X_{t+1}|X_t) \tag{7.3}$$

and

$$H(X_{t+1}|\mathbf{x}_t) = -\sum_{\mathbf{x} \in \mathbf{X}} p(\mathbf{x}) \sum_{X_{t+1} \in \mathbf{X}} p(X_{t+1}|\mathbf{x}_t) \log p(X_{t+1}|\mathbf{x}_t). \tag{7.4}$$

Here $p(X_t)$ is the stationary probability that the process is in the macro state X_t computed on the basis of microscopic stationary probabilities as $p(X_t) = \sum_{\mathbf{x} \in X_t} p(\mathbf{x})$ and for the strongly lumpable case the probabilities $p(\mathbf{x})$ are equal for all micro states in the same macroscopic set. Likewise, strong lumpability requires that the probabilities $p(X_{t+1}|\mathbf{x}_t)$ are equal for all micro states in the same macroscopic state (X_t). For this reason, the conditional entropies $H(X_{t+1}|\mathbf{x}_t)$ are maximal and equal to $H(X_{t+1}|X_t)$ so that the information flow vanishes.

However, for the case of weak lumpability the situation is not so simple. Namely because weak lumpability does not discard the case of heterogeneous probabilities ($p(X_{n+1}|\mathbf{x}) \neq p(X_{n+1}|\mathbf{y})$; $\mathbf{x}, \mathbf{y} \in X_n$) such that the transition probability to the set X_{n+1} may be high for one but low for another micro state within the same subset X_n. Intuitively, therefore, knowledge of the micro state (whether the process is in $\mathbf{x}$ or $\mathbf{y}$) may provide additional information about the transition to X_{n+1}.

In fact, both things may happen: there are cases of weak lumpability in which informational flow vanishes and others where it does not. An example for the former is the VM on the chain (or the ring). The corresponding micro chain is weakly lumpable with respect to the macroscopic partition $\mathbf{X}$ but information flow measured by Eq. (7.2) vanishes.[1] An example for the situation that information flow does not

---

[1]This example will be explored carefully in future work. For now, just notice that the VM on the ring leads to a long-lasting pattern of a single white and a single black regime and further change in the number of white and black agents only happens when an edge at the interface between the two regimes is chosen. The respective probability is equal for all micro configurations of this kind and other (disordered) configurations are not visited once such a situation has been reached. See also Fig. 1.1 in the Introduction.

vanish is provided with the weakly lumpable example chain given in Kemeny and Snell (1976, Ex. 6.4.2)

$$
\hat{P} = \begin{array}{c} S_1 \\ \\ S_2 \end{array} \begin{pmatrix} \frac{1}{4} & \frac{1}{4} & \frac{1}{2} \\ 0 & \frac{1}{6} & \frac{5}{6} \\ \frac{7}{8} & \frac{1}{8} & 0 \end{pmatrix} \xrightarrow{(1-3a,a,2a)} P = \begin{array}{c} S_1 \\ S_2 \end{array} \begin{pmatrix} \frac{1}{4} & \frac{3}{4} \\ \frac{7}{12} & \frac{5}{12} \end{pmatrix}. \tag{7.5}
$$

This chain is weakly lumpable with respect to the partition (S_1, S_2) for all vectors of the form $(1 - 3a, a, 2a)$ and, in particular, for the stationary vector $(7/16, 3/16, 3/8)$ leading to the chain P on the right-hand side. The information flow $(I(X_{n+1} : \mathbf{x}_n | X_n))$ for this example is $9/16 \log(3) - 15/64 \log(5) \approx 0.347$.

Finally, as noted in the previous section, Pfante et al. (2014a) shows that $I(X_{n+1} : \mathbf{x}_n | X_n) \geq I(X_{n+1} : X_{-\infty}^{n-1} | X_n)$ so that vanishing information flow from micro to macro implies Markovianity. For strong lumpability both measures vanish. For weak lumpability, information flow may in fact only vanish in special cases (such as the VM on the ring).

7.2 Network Dynamics from the Macro Perspective

We now turn to the CVM and consider the behavior on different networks. We first recall what we consider as the macro level throughout this chapter.

7.2.1 Full Aggregation

The most natural level of observation in binary state dynamics is to consider the temporal evolution of the attribute densities, or respectively, the number of agents in the two different states. While a mean-field description would typically formulate the macro dynamics a differential equation describing the evolution of attribute densities, the Markov chain approach operates with a discrete description (in time as well as in space) in which all possible levels of absolute attribute frequencies and transitions between them are taken into account. Regardless of the microscopic details such as more complex interaction networks or rules, a macro level description of that kind—always a tremendous reduction of original system— is desirable in order to obtain a better understanding of the model behavior. As a matter of fact, it is desirable for both numerical as well as analytical arguments.

One of the main contributions of the framework proposed here is that the link between the microscopic system and a certain macro level description is made explicit. In order to better understand the Markovian as well as the non-Markovian cases such an explicit link between the micro and the macro level is a prerequisite. Let us, as before, denote as k the number of $\square$-agents in the population ($k = N_\square$)

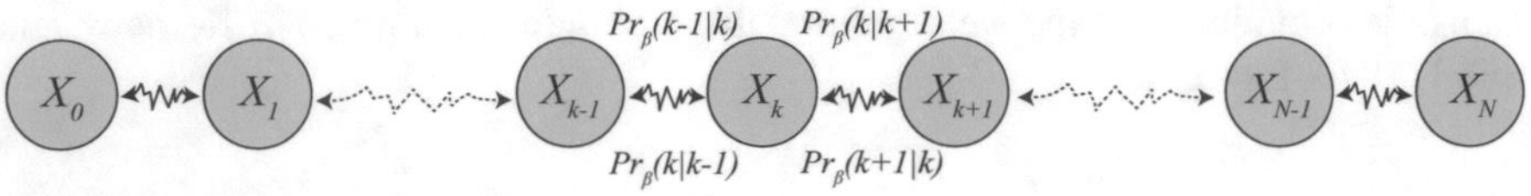

Fig. 7.2 Full aggregation is obtained by the agglomeration of states with the same Hamming weight $h(\mathbf{x}) = k$. The resulting macro process is, in general, a non-Markovian process on the line

and refer to this level of observation as global or *full aggregation*. As explained in detail in Chap. 3, Sect. 3.3, the system-level property $k = h(\mathbf{x})$ induces a partition $\mathbf{X} = \{X_0, \dots, X_k, \dots, X_N\}$, with $0 \leq k \leq N$ on the space of all possible micro configurations and the macro-level process corresponds to the original micro chain $(\Sigma, \hat{P})$ projected onto that partition. Notice again that in case of the CVM the micro process is a random walk on the N-dimensional hypercube and that each macro state X_k collects all micro configurations with the same Hamming weight.

In this regard, one important observation in Chap. 3 has been that, for the VM and related models, homogeneous mixing is a prerequisite for lumpability, and that microscopic heterogeneities (be it in the agents or in their connections) translate into dynamical irregularities that prevent lumpability with respect to $\mathbf{X}$. This means that full aggregation over the agent attributes $(k = h(\mathbf{x}))$ leads in general to a non-Markovian macro process. We illustrate this process in Fig. 7.2. Still, the process obtained by the projection from micro to macro is characterized by the fact that from an atom X_k the only possible transitions are the loop to X_k, or a transition to neighboring atoms X_{k-1} and X_{k+1}. This is due to the fact that the CVM implements single-step dynamics in which only one agent changes at a time. However, the micro level transition rates—see Eq. (6.2) in the previous chapter— depend essentially on the connectivity structure between the agents, and therefore, the transition probabilities at the macro level (denoted as $Pr_\beta(l|k)$ in Fig. 7.2) are not uniquely defined (except for the case of homogeneous mixing). That is, for two configurations in the same macro state $\mathbf{x}, \mathbf{x}' \in X_k$ the probability to go to another macro state (e.g., X_{k+1}) may be very different which violates the lumpability conditions of Theorem 6.3.2 in Kemeny and Snell (1976).

The information-theoretic measures introduced above provide us with instruments to study the macro-level effects to which a non-trivial interaction structure at the micro level may lead. The questions we aim to address are of the following type: Why and in what sense does the behavior of the macro process deviate from Markovianity? Do we introduce memory or even long-range correlations at the macro level by the very way we observe the process? Is the emergence of these effects just due to an aggregation which is insensitive to microscopic heterogeneities? And furthermore: How good does the mean field (homogeneous mixing) solution approximate network dynamics and for which networks does it provide acceptable approximations? Is there an alternative assignment of probabilities $Pr_\beta(l|k)$ that leads to better results? Which properties can be captured? Finally, an interesting question concerns the reducibility of the micro chain by the weaker

form of lumpability. At least to some of these question answers will be provided in
the remainder of this chapter.

7.2.2   Network Influence on the Stationary Dynamics

Let us first consider, in a numerical experiment, the effect of different interaction
topologies ω on the stationary dynamics of the resulting macro process. For this
purpose, we define the stationary macro measure π as

$$\pi_k = \sum_{\mathbf{x} \in X_k} \hat{\pi}_{\mathbf{x}}. \tag{7.6}$$

In other words, the elements π_k of the stationary vector are determined by counting
the frequency with which the model is in the respective set of micro states with
$h(\mathbf{x}) = k$. Notice that on the basis of a stationary micro chain, it is always possible
to construct an approximate macro chain—an aggregation—the stationary vector of
which satisfies Eq. (7.6) (see Kemeny and Snell 1976, p. 140 and Buchholz 1994,
pp. 61–63). This will be discussed below.

To compute the π_k, a series of simulations has been performed in which the
CVM with $N = 100$ is run on different paradigmatic agent networks. To capture
the model in stationarity, the model is iterated for several thousands of steps first
and the statistics of this "burn-in" phase are not considered in the computation of
π_k. (In this exploratory analysis with 100 agents a "burn-in" period of 20,000 steps
has been used.) The result is shown in Fig. 7.3 for the case of a small contrarian rate
$p = 0.005$.

We observe in Fig. 7.3 that some interaction topologies give rise to strong
deviations from the theoretical result derived for homogeneous mixing (solid, blue).
In general, there is an increase in the probability to observe balanced configurations
and the case of complete consensus tends to become less likely. However, the
results obtained for the random graph are indeed very similar to the theoretical
prediction and also the scale-free topology leads to stationary statistics that, in
qualitative terms, correspond to the mean-field case. On the other hand, we observe
a strong "modulation" of the stationary statistics by networks that tend to foster
the emergence of "local alignment and global polarization". By local alignment
and the dynamics that lead to it, we refer to situations in which different clusters
of agents approach independently a certain local consensus which is in general
different from agent cluster to agent cluster. From the global perspective the entire
population appears to be far from complete consensus and the probability to observe
the respective intermediate macro states is increased. These effects are observed for
the small-world network, the two-community graph as well as for the lattice, and it
is strongest for the ring where the probability of complete consensus is practically
zero.

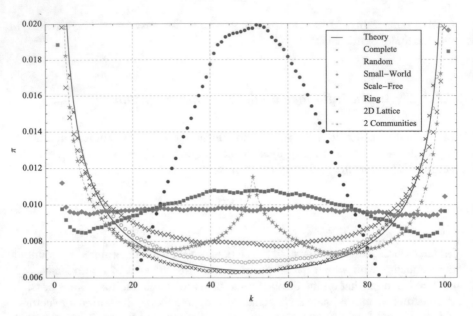

Fig. 7.3 Stationary statistics for the CVM on different topologies. Due to effects of local ordering, the stationary behavior of the small-world network, the ring and the lattice as well as the two-community topology differs greatly from the well-mixed situations

7.2.3   The Two-Community Case

For the two-community graph with a peak around $k = N/2$ the interpretation of the result is particularly straightforward. Local alignment, in this case, refers to inter-community polarization—the situation in which a different consensus has emerged in the two communities. If the size of the communities is $N/2$, as in the example we study, the polarization configurations give rise to an macro observation $k = N/2$ since one half of the population (organized in one community) agrees on □ and the other half (that is, the other community) is in state ■.

The two-community CVM is particularly interesting because we can compute the exact stationary vector by analyzing the respective meso chain $(\tilde{\mathbf{X}}, \tilde{P})$ obtained via strong lumpability. This has been done in Sect. 6.4.2. We first compute the stationary distribution of the meso chain assigning the respective limiting probability $\tilde{\pi}_{m,l}$ to each state $\tilde{X}_{m,l}$. In that case, Eq. 7.6 reads

$$\pi_k = \sum_{m+l=k} \tilde{\pi}_{m,l}. \tag{7.7}$$

That is, π_k associated to the macro state X_k is obtained by summing up the respective $\tilde{\pi}_{m,l}$ with $m + l = k$. This is shown in Fig. 7.4 for different contrarian rates p and different couplings between the two sub-graphs.

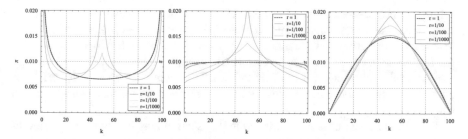

Fig. 7.4 The stationary distribution from the macro perspective for different $r = \alpha/\gamma = 1, 1/10, 1/100, 1/1000$. From *left* to *right* $p = 0.005, 0.01, 0.02$

It becomes clear that the probability to observe a fifty-fifty situation ($k \approx N/2$) generally increases, the weaker the coupling between the communities. The analysis of the meso level stationary distribution shown in Fig. 6.7 (previous chapter) makes clear that this is due to an increased probability for the configurations with intra-community consensus and inter-community polarization ($\tilde{X}_{N/2,0}, \tilde{X}_{0,N/2}$) which contribute to that probability. (Notice that the community sizes have a direct effect onto the macro level statistics and that in general the states with $k = L$ and $k = M$ will be observed more frequently when the coupling is weak.) In general, we can also observe that the influence of the different topological choices onto the macro behavior (captured here in terms of π_k) decreases with an increasing contrarian rate p. As explained in Sect. 6.4.2, the more contrarian behavior is allowed by the parameter setting, the more random becomes the entire process which undermines the effects of local alignment and, consequently, of interaction topology. This can be taken is a first indication that the mean-field solution (here represented by $r = 1$) might approximate well the model behavior with a relatively high contrarian rate because the entire setting is characterized more and more by random state flips. It will be less accurate for a small contrarian rate where dynamics of local ordering become more and more characteristic.

7.3 Non-Markovianity of the Two-Community CVM

7.3.1 From Micro to Meso, and from Meso to Macro

The previous section has shown that heterogeneous interaction structures can have a strong impact on the model behavior. From the lumpability point of view, but also from the point of view of observation, a macro process obtained by global aggregation over the agent attributes neglects important information about the microscopic details. In other words, for heterogeneous networks the macroscopic process describing the dynamics of the system by the state variables associated to the level of full aggregation does not provide a "closed" description of the original micro process derived from the ABM. A series of measures—among them conditional past-future mutual information (Görnerup and Jacobi 2008) and micro-to-macro information flow (Pfante et al. 2014a)—have been developed to quantify the deviations from such an idealized closed description (see Sect. 2.4 and the first section of this chapter for a brief overview). The remainder of this chapter is an attempt to better understand this loss of information and the macro-level effects this leads to.

Even though the questions addressed in this section may be not directly relevant for an interpretation in terms of a specific application (such as opinion dynamics in case of the CVM), an improved understanding of the dynamical effects introduced by the way an agent system is observed is of great relevance for ABMs more generally. The simple rules of the CVM along with the controllable two-community topology make this scenario well-suited for a first step to analyze the effects at an aggregate level introduced by aggregation without sensitivity to micro- or mesoscopic structures.

The general idea is illustrated in Fig. 7.5. Consider the CVM on the two-community graph and the associated micro-level process $(\Sigma, \hat{P})$. The two-community micro chain $(\Sigma, \hat{P})$ is (strongly) lumpable with respect to the partition $\tilde{\mathbf{X}}$. This gives rise to what we have called the meso-level process $(\tilde{\mathbf{X}}, \tilde{P})$ in the previous chapter (see Sect. 6.4). The meso chain gives us a complete understanding of the (micro) behavior of the CVM on two coupled communities, because the coarse-graining via strong lumpability is compatible with the exact symmetries of the micro process. That is, no information is lost by a formulation of the dynamics in terms of the frequencies m and l in the two communities. However, the process (the micro as well as the meso chain) is not lumpable with respect to the macro level of

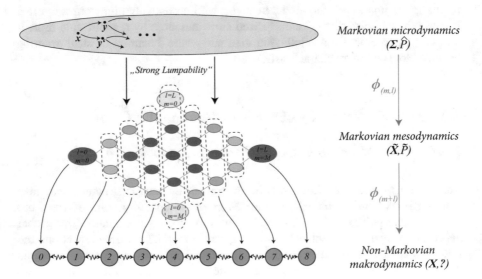

Fig. 7.5 From micro to meso, and from meso to macro

full aggregation (partition $\mathbf{X}$) which formulates the dynamics in terms of the opinion frequency in the entire population ($k = m + l$). Therefore, if we wish to observe the process at the global level, which is often the case in binary state models, we must live with the fact that the resulting macro process on $\mathbf{X}$ is no longer a Markov chain. In other words, more complex temporal correlations (memory effects) emerge at the macro level.

As illustrated in Fig. 7.5, here we project onto the level of full aggregation despite the fact that Markovianity is lost, in order to understand (1) the reasons for which lumpability is violated and (2) the dynamical effects that this introduces at the macro level. That is, all meso states $\tilde{X}_{m,l}$ with the same global opinion frequency $k = m + l$ are projected into the same macro state X_k. We make use of the fact the two-community coarse-graining ($\tilde{\mathbf{X}}$) is a proper refinement of the full aggregation ($\mathbf{X}$) which describes exactly the model dynamics on an interaction topology with a small amount of inhomogeneity and is, at the same time, small enough for explicit computations. This explicit understanding of the meso chain facilitates an explicit analysis of the transition from micro to meso to macro.

7.3.2 Why Lumpability Fails

Let us first inspect the reasons for which the meso chain ($\tilde{\mathbf{X}}, \tilde{P}$) is not lumpable with respect to the macro partition $\mathbf{X}$. By the lumpability theorem (Kemeny and Snell 1976, Theorem 6.3.2), it is clear that the non lumpability of the meso chain with respect to $\mathbf{X}$ comes by the fact that the transition probabilities $Pr(X_k|\tilde{X}_{m,l})$ are not equal for all meso states $\tilde{X}_{m,l} \in X_{(m+l)}$ in the same macro set. As an example, let us consider the transition rates from the single $\tilde{X}_{m,l} \in X_{50}$ to the macro set X_{51} in a system with $M = L = 50$. One by one, the conjoint probability from $\tilde{X}_{0,50}, \tilde{X}_{1,49}, \tilde{X}_{2,48}, \ldots, \tilde{X}_{50,0}$ to the sets $\tilde{X}_{m,l} \in X_{51}$ is shown in Fig. 7.6 for various ratios r and a small contrarian rate $p = 0.01$.

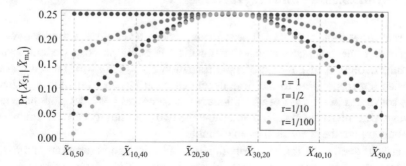

Fig. 7.6 The island topology leads to inhomogeneous transition probabilities and is therefore not (strongly) lumpable. Here the example of a transition from $\tilde{X}_{m,l} \in X_{50}$ to X_{51} in a system with $M = L = 50$ is shown

We first notice that the transition rates $\tilde{P}(\tilde{X}_{m,l}, X_{(m+l+1)})$ are uniform when the coupling within is equal to the coupling across communities, that is for $\alpha = \gamma$ and $r = 1$. Obviously, this is the case of homogeneous mixing and the uniformity of the $\tilde{P}(\tilde{X}_{m,l}, X_{(m+l+1)})$ is precisely the lumpability condition spelled out in Theorem 6.3.2 of Kemeny and Snell (1976).

In general, the $\tilde{P}(\tilde{X}_{m,l}, X_{(m+l+1)})$ are no longer equal for all m and l with $m+l = k$ when heterogeneity is introduced in form of a different coupling within and across communities, i.e., $\alpha \neq \gamma$. This explains the non-lumpability of the two-community model with respect to $\mathbf{X}$. As the weak ties across communities becomes weaker such that the ratio r between strong and weak ties decreases, the transition rates become inhomogeneous, the main effect being a strong decrease of $\tilde{P}(\tilde{X}_{m,l}, X_{(m+l+1)})$ for the atoms close to polarization ($m = M, l = 0$ and $m = 0, l = L$). This decrease in transition probability, in turn, explains the increased stationary probability of the states $\tilde{X}_{50,0}$ and $\tilde{X}_{0,50}$, because once entered there is a relatively small probability to leave them so that the process is likely to "wait" in these states for quite some time.

Notice that there is only the small difference in transition rates between $r = 1/10$ and $r = 1/100$ (the difference to $r = 1/1000$ is even smaller!). On the one hand, this is somewhat surprising, as from the dynamical point of view $r = 1/10$ is much more related to the homogeneous mixing case ($r = 1$) than to the situation with $r = 1/100$ (cf. Fig. 7.4). On the other hand, the probability to leave a polarized state ($\tilde{X}_{0,50}, \tilde{X}_{50,0}$) decreases significantly with every decrease in r and therefore the waiting times for these states grow tremendously. Notice, however, that in the limit of $r \to 0$, the probability of leaving a polarized state converges to p (with $\tilde{P}(\tilde{X}_{M,0}, X_{(M+1)}) = \tilde{P}(\tilde{X}_{M,0}, X_{(M-1)}) = p/2$). Therefore a strong difference between a weak (e.g., $r = 1/100$) and a very weak coupling ($r = 1/1000$) in form of an increased stationary probability of polarization can be expected only if also the contrarian rate p is small. Likewise, as already observed in Sect. 6.4.2, a large contrarian rate can completely undermine effects of polarization altogether.

7.3.3 Stationarity and Aggregation

We shall now look at what happens to the macro level system as the micro or respectively meso process reaches stationarity. For this purpose we first look at the time evolution of the macroscopic transition rates. It is well-known that this measure (corresponding to the time dependent distribution over blocks of length two) converges in the case of an stationary macro process. We develop these ideas for a general micro chain $(\Sigma, \hat{P})$ and show the two-community case (where we can indeed compute these entities) as an example.

Let $\hat{\beta}(0)$ denote the initial distribution over all micro configurations and $\hat{\beta}(t)$ be the respective distribution at time t. Notice that $\hat{\beta}(t) = \hat{\beta}(0)\hat{P}^t$. Let us further define the probability distribution at time t restricted to the macro set $X_k \in \mathbf{X}$ as $\hat{\beta}^k(t)$. That is, the $\mathbf{x}$th element $\hat{\beta}_{\mathbf{x}}^k(t) = 0$ whenever $\mathbf{x} \notin X_k$ and proportional to $\hat{\beta}_{\mathbf{x}}(t)$ with

$$\hat{\beta}_{\mathbf{x}}^k(t) = \frac{\hat{\beta}_{\mathbf{x}}(t)}{\sum\limits_{\forall \mathbf{x}' \in X_k} \hat{\beta}_{\mathbf{x}'}(t)}, \tag{7.8}$$

for every $\mathbf{x} \in X_k$. Notice that by convention $\hat{\beta}_{\mathbf{x}}^k(t) = 0$ whenever $\mathbf{x} \notin X_k$ and that $\hat{\beta}^k(t)$ is defined only if $\sum\limits_{\mathbf{x}' \in X_k} \hat{\beta}_{\mathbf{x}'}(t) > 0$, that is, if there is a positive probability that the process has reached at least one configuration $\mathbf{x}' \in X_k$. The probability $\hat{\beta}_{\mathbf{x}}^k(t)$ shall be interpreted as the conditional probability that the process is in the configuration $\mathbf{x}$ at time t provided that it is in the set X_k at that time.

We now denote the expected transition probability from macro state X_k to macro state X_s as $Pr_{\hat{\beta}(0)}^t(X_s|X_k)$. With $\hat{\beta}^k(t)$ defined as above, it is given by

$$Pr_{\hat{\beta}(0)}^t(X_s|X_k) = \sum_{\mathbf{x} \in X_k} \left[\hat{\beta}_{\mathbf{x}}^k(t) \sum_{\mathbf{y} \in X_s} \hat{P}(\mathbf{x}, \mathbf{y}) \right]. \tag{7.9}$$

For the interpretation of Eq. (7.9) consider that $\hat{\beta}_{\mathbf{x}}^k(t)$ is the probability (restricted to X_k) that the process is in $\mathbf{x} \in X_k$ at time t and $\sum\limits_{\mathbf{y} \in X_s} \hat{P}(\mathbf{x}, \mathbf{y}) = \hat{P}(\mathbf{x}, X_s)$ is the probability for a transition from $\mathbf{x}$ to some $\mathbf{y} \in X_s$. A transition from the set X_k to X_s is then the conjoint transition probability considering all $\mathbf{x} \in X_k$ along with their conditional probability $\hat{\beta}_{\mathbf{x}}^k(t)$ (first sum). Notice again that (7.9) corresponds to the probability of observing a sequence of two measurements $(h(\mathbf{x}), h(\mathbf{y})) = (k, s)$ at a certain time t when looking at the micro system through the eye of absolute attribute frequencies.

Notice also that we can write Eq. (7.9) in matrix form as

$$P = U_\beta \hat{P} V, \tag{7.10}$$

$\hat{P}$ being the transition matrix of the original process and U_β and V define the projection ϕ as follows: The number of states of the original chain $(\Sigma, \hat{P})$ and the reduced macro process $\mathbf{X}$ is 2^N and $N + 1$ respectively ($N + 1 < 2^N$). Then V is a $2^N \times N + 1$ matrix with $V_{\mathbf{x}k} = 1$ if the micro state $\mathbf{x}$ is mapped to the macro state X_k by ϕ and zero elsewhere (i.e., if $\phi(\mathbf{x}) = k$). U is a $N + 1 \times 2^N$ matrix defined in a similar way such that $U_{k\mathbf{x}} > 0$ whenever $\phi(\mathbf{x}) = X_k$. However, the values of $U_{k\mathbf{x}}$ are chosen such that information about the distribution β is included. Namely, $U_{k\mathbf{x}} = \hat{\beta}_{\mathbf{x}}^k$ (7.8).

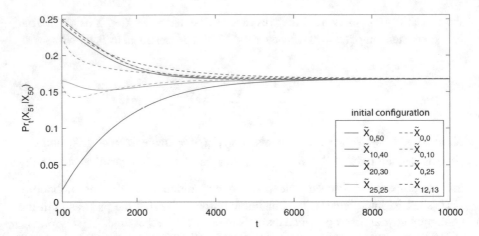

Fig. 7.7 Time evolution of transition rates $Pr^t_{\hat{\beta}(0)}(X_{51}|X_{50})$ from X_{50} to X_{51} in the two-community model for some of the meso states considered in Fig. 7.6 and different initial conditions

Now, notice that the only time dependent term in Eq. (7.9) is the conditional distribution $\hat{\beta}^k_{\mathbf{x}}(t)$ which is obtain by (7.8) from $\hat{\beta}(t)$, and $\hat{\beta}(t) = \hat{\beta}(0)\hat{P}^t$. Considering that $(\Sigma, \hat{P})$ is regular, it is clear that the process reaches its stationary state ($\lim_{t\to\infty} \hat{\beta}(t) = \hat{\pi}$) independent of the initial $\hat{\beta}(0)$. Therefore, the $Pr^t_{\hat{\beta}(0)}(X_s|X_k)$ converge to

$$Pr_{\hat{\pi}}(X_s|X_k) = \sum_{\mathbf{x}\in X_k}\left[\hat{\pi}^k_{\mathbf{x}}\sum_{\mathbf{y}\in X_s}\hat{P}(\mathbf{x},\mathbf{y})\right] \qquad (7.11)$$

as the micro process reaches stationarity. See Fig. 7.7 for the two community model. Consequently (Kemeny and Snell 1976; Buchholz 1994), Eq. (7.11) can be interpreted as a macroscopic transition matrix with $P(X_k, X_s) = Pr_{\hat{\pi}}(X_s|X_k)$, and the stationary vector of that matrix will be correct in the sense of Eq. (7.6).

The possibility of deriving such a macro description has been commented on by Kemeny and Snell (1976), p. 140, and it is discussed with some detail by Buchholz (1994), pp. 61–63, where it is referred to as an *ideal aggregate*. The most important thing to notice (Kemeny and Snell 1976, p. 140) is that P^2 does not correctly describe the two-step transition probabilities that would be measured on the micro system. That is, the system evolution described solely at the aggregated macro level is different from the macro evolution that would be observed by running the microscopic process and performing an aggregation after each micro step. In other words, in the general non-lumpable case we have $(U_\pi \hat{P}V)(U_\pi \hat{P}V) \neq (U_\pi \hat{P}\hat{P}V)$. The reader may be referred to Kemeny and Snell (1976, pp. 135/136) where this is used to derive conditions for (weak) lumpability. In fact, one can basically look at an ideal aggregate obtained by (7.11) as a Markov model that approximates a certain stationary process (in our case the macro process obtained by measurements from the micro chain) on the basis of the empirical distribution of cylinders of length two. It is in fact not clear whether the process is informative about certain properties of the real macro process beyond the stationary measure (see Sect. 2.3.3). Finally, even if the chain defined by (7.11) would be informative about certain transient properties of the real macro process, it still suffers from the fact that the construction of it requires knowledge of the stationary distribution of the micro chain $\hat{\pi}$ which is usually unknown.

Notice that Eqs. (7.9) and (7.11) do not involve any particular assumption on the nature of the partition meaning that an ideal aggregate can be constructed by them for any partition of Σ. Buchholz (1994), Theorem 1, has shown that if the original transition matrix ($\hat{P}$ in our case) is irreducible than the transition matrix of the ideal aggregate $P(X_k, X_s) = Pr_{\hat{\pi}}(X_s | X_k)$ will also be irreducible and therefore possess a unique stationary distribution.

7.3.4 Why Weak Lumpability Fails

Weak lumpability (see Sect. 2.3.2) refers to the fact that a Markov chain might be lumpable only for particular starting vectors (Burke and Rosenblatt 1958; Kemeny and Snell 1976; Ledoux et al. 1994). The question whether or not an ideal aggregate (and hence the micro chain) is weakly lumpable arises naturally from our construction of an ideal aggregate, (7.9) and (7.11), mainly by two considerations: first, it is well-known that if a chain is weakly lumpable with respect to some distribution, it must be lumpable with respect to the stationary distribution; and second, the transition probabilities of the lumped process would be given by Eq. (7.11) (Kemeny and Snell 1976, Theorem 6.4.3). Therefore questions of weak lumpability of the micro process with respect to full aggregation $\mathbf{X}$ can be answered by checking if the ideal aggregate is lumpable.

For the two-community model it is in fact easy to show that the CVM process is not weakly lumpable by the construction of a counter example which shows that the conditions of Theorem 6.4.1 in Kemeny and Snell (1976) are violated. The argument is twofold. First, starting from $\hat{\pi}$ the process generally reaches different assignments of probabilities over the micro states in the different macro sets (different $\hat{\beta}^s$), because, at least for the two-community model,

$$(\pi^k P)^s \neq \pi^s. \tag{7.12}$$

The superscripts k and s denotes, as before, restriction to X_k and X_s respectively. Let us denote the left-hand side of (7.12) as $\hat{\pi}'^s = (\pi^k P)^s$. Notice that, in fact, for weak lumpability it would be sufficient to show that $\hat{\pi}'^s = \pi^s$ is satisfied for any k and s (cf. Kemeny and Snell 1976, p.136). However, even if the situation is as in (7.12), weak lumpability could still be the case if the two distribution $\hat{\pi}'^s$ and π^s lead to the same transition probabilities to all other macro sets X_l

$$Pr_{\hat{\pi}}(X_l|X_s) = Pr_{\hat{\pi}'}(X_l|X_s). \tag{7.13}$$

In other words, weak lumpability (according to Kemeny and Snell 1976, Theorem 6.4.1) is violated if the probability of a transition from X_s to another macro state X_l is different for $\hat{\pi}^s$ and $\hat{\pi}'^s$. This is the case for the two-community model, as will be shown in the sequel.

As an example, let us consider a small system with $M = L = 2$. That is, the two communities each consist of only two agents. Let us say the process is in equilibrium with distribution $\hat{\pi}$ at time t. Now we consider the macro probability $X_2 \rightarrow X_1$, $Pr_{\hat{\pi}}(X_1|X_2)$, which is given by:

$$Pr_{\hat{\pi}}(X_1|X_2) = \frac{(1 + r - p)(2r(-1 + p) - p)(1 + p)}{2(-1 + 2r^2(-2 + p) + 2p - 3p^2 + r(-1 - 7p + 6p^2))} \tag{7.14}$$

for arbitrary r and p. Let us further assume that the process performs a loop in the first step ($t \rightarrow t + 1$) and transits to X_1 only after that (in $t + 1 \rightarrow t + 2$). That is, $X_2 \rightarrow X_2 \rightarrow X_1$. For weak lumpability with starting vector $\hat{\pi}$ the probability of $X_2 \rightarrow X_1$ must be the same independent of how many and which previous steps are taken. However, for the second case we have $\hat{\pi}'^2 = (\pi^2 P)^2 \neq \pi^2$ and then

$$Pr_{\hat{\pi}'}(X_1|X_2) = \frac{(1 + r - p)(2r(-1 + p) - p)\left(1 - 2p - 4r(-2 + p)p + 3p^2 + r^2\left(2 + 4p^2\right)\right)}{2(1 + 2r)^2\left(1 - 3p + 3p^2 + p^3 + 2r^2\left(1 - p + p^2\right) - r\left(1 - 8p + 5p^2 + 2p^3\right)\right)}, \tag{7.15}$$

which is obviously not equal to (7.14). This shows that the two-community model is not weakly lumpable with respect to $\mathbf{X}$.

In Fig. 7.8 we show the probabilities $Pr_{\hat{\beta}}(X_1|X_2)$ for the cases from $X_2 \rightarrow X_1$ to $X_2 \rightarrow X_2 \rightarrow X_2 \rightarrow X_2 \rightarrow X_1$ as a function of p (top) and r (bottom). As we would

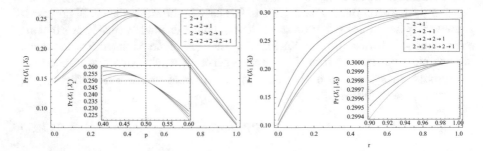

Fig. 7.8 Transition probabilities $Pr_{\hat{\beta}}(X_1|X_2)$ for $\hat{\beta} = \hat{\pi}, \hat{\pi}', \hat{\pi}'', \hat{\pi}'''$ for the small example $M = L = 2$ are not equal as would be required for weak lumpability. *Top*: $Pr_{\hat{\beta}}(X_1|X_2)$ is shown as a function of p for $r = 1/5$. The *curves* converge to the same value at $p = 1/2$. *Bottom*: $Pr_{\hat{\beta}}(X_1|X_2)$ is shown as a function of r for $p = 1/5$. Equal probabilities are observed for the strongly lumpable case $r = 1$

expect (see figure on the bottom and the inset) the curves approach the same value as $r \to 1$. This is the strongly lumpable case of homogeneous mixing. Interestingly, we observe in the upper image of Fig. 7.8 that the probabilities are actually equal for $p = 1/2$, namely $Pr_{\hat{\beta}}(X_1|X_2) = 1/4$ in that case. See the respective inset in the upper figure. This indicates lumpability of the process for $p = 1/2$ and, in fact, it is possible to show that the two-community model is strongly lumpable whenever $p = 1/2$. The reason is that for $p = 1/2$, the meso-level transition matrix $\tilde{P}$ is independent of the topological parameter r. Even if the case $p = 1/2$ is not that interesting from the point of view of the dynamical behavior of the CVM, it would be interesting to check whether a similar effect also occurs for other networks.

7.4 Closure Measures for the Two-Community CVM

7.4.1 Computation of the Markovianity Measure

Having shown that the macro process associated to the CVM on two coupled communities is non-Markovian, the next logical step is to quantify in some way the deviations from Markovianity. The framework of information theory—relative entropy and mutual information in particular—has been shown to be quite useful for this purpose (Chazottes et al. 1998; Vilela Mendes et al. 2002; Görnerup and Jacobi 2008; Ball et al. 2010; James et al. 2011; Pfante et al. 2014a, among others). Here we motivate the Markovianity measure introduced previously as the mutual information $I(X_{n+1} : X_{-\infty}^{n-1}|X_n)$ (see Chap. 2 and Sect. 7.1 of this chapter) from the perspective of time series and dynamical systems.

Let us, to simplify the writing, denote as $[\ldots, k_{t-2}, k_{t-1}, k_t, k_{t+1}, k_{t+2,\ldots}]$ a sequence of macro states $\ldots \to X_{k_{t-2}} \to X_{k_{t-1}} \to X_{k_t} \to X_{k_{t+1}} \to X_{k_{t+2}} \to \ldots$. Likewise, let us denote as $[k_{t-m}, \ldots, k_t]$ a finite sequence of m macro states and refer to this as block or cylinder of length m. Then, the block entropy associated to cylinders of length m is defined by

$$H_m = - \sum_{[k_{t-m},\ldots,k_t] \in \mathfrak{G}_m} \mu([k_{t-m}, \ldots, k_t]) \log \mu([k_{t-m}, \ldots, k_t]) \tag{7.16}$$

where $\mu([k_{t-m}, \ldots, k_t])$ denotes the probability to observe the respective cylinder $[k_{t-m}, \ldots, k_t]$. Notice that for $m > 1$ there exist in general "forbidden" sequences with $\mu([k_{t-m}, \ldots, k_t]) = 0$, a fact that is usually formalized in terms of a grammar $\mathfrak{G}_m \subseteq \mathbf{X}^m$ by defining $\mathfrak{G}_m := \{[k_{t-m}, \ldots, k_t] : \mu([k_{t-m}, \ldots, k_t]) > 0\}$. In our case of single-step dynamics, all sequences containing subsequent elements with $|k_t - k_{t-1}| > 1$ are "forbidden" because only X_k, X_{k-1} and X_{k+1} can be reached from X_k in one step.

It is well-known (Chazottes et al. 1998; Vilela Mendes et al. 2002; James et al. 2011) that the slope of the block entropy $\Delta H_m = H_m - H_{m-1}$ converges to a fixed value called entropy rate (usually denoted as $h(\mu)$) and that this fact can be used to estimate the memory range of the process. Namely, following Chazottes et al. (1998) and Vilela Mendes et al. (2002), the range of the process is given, at least in an approximative sense, by the m at which ΔH_m reaches a constant value, that is, $\Delta H_m - \Delta H_{m+1} \approx 0$. It is clear then that for a Markovian process this point must be reached at $m = 2$ such that

$$\Delta H_2 - \Delta H_3 = 0 \tag{7.17}$$

and more generally

$$\Delta H_2 - \Delta H_m = 0. \tag{7.18}$$

Notice that Eq. (7.18) is precisely the "Markov property measure" proposed in Görnerup and Jacobi (2008), pp.6–8, to identify projections of a process onto a smaller state space (a partition of the original process) which lead to Markovian dynamics. Noteworthy, the starting point of Görnerup and Jacobi (2008) and consequently of Pfante et al. (2014a) has been the expected mutual information $\langle I \rangle$ between pasts and the future state. The aim is to quantify how much information about the next symbol (k_{t+1}) is on average over all symbols contained in the sequence of symbols ($[\dots, k_{t-2}, k_{t-1}]$) before the current symbol (k_t), or, in other words, how much does knowledge of the past reduce uncertainty of k_{t+1} given k_t. We follow Görnerup and Jacobi (2008) and Pfante et al. (2014a) and write

$$\begin{aligned}\langle I \rangle &= I(k_{t+1} : k_{[-\infty,\dots,t-1]}|k_t) \\ &= H(k_{t+1}|k_t) - H(k_{t+1}|k_{[-\infty,\dots,t-1]}) \\ &= \Delta H_2 - \Delta H_\infty \end{aligned} \tag{7.19}$$

which shows that the expected past future mutual information $I(k_{t+1} : k_{[-\infty,\dots,t-1]}|k_t)$ can be expressed in terms of the slopes of the block entropy relating it directly to the previous considerations.

In real computations, however, one always has to restrict to finite histories since it is not possible to compute $\Delta H_\infty = H(k_{t+1}|k_{[-\infty,\dots,t-1]})$ in practice. Therefore, Görnerup and Jacobi (2008) introduce the finite history variant

$$\langle I_n \rangle = I(k_{t+1} : k_{[t-n,\dots,t-1]}|k_t) = \Delta H_2 - \Delta H_{2+n} \tag{7.20}$$

and compute $\langle I_2 \rangle = \Delta H_2 - \Delta H_4$ for their examples which means that they have to consider cylinders up to length four $[k_{t-2}, k_{t-1}, k_t, k_{t+1}]$ in their Markovianity test. Notice that in their notation n accounts for the ranges beyond the Markov range of two ($m = n + 2$ in Eq. 7.18). We will follow this notation here and compute

Fig. 7.9 Possible paths of
length 3 through X_k

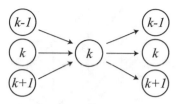

$\langle I_1 \rangle = \Delta H_2 - \Delta H_3$ and $\langle I_2 \rangle = \Delta H_2 - \Delta H_4$, the latter being used also by Görnerup and Jacobi (2008).

The advantage of the two-community CVM as a framework to link between a micro and a macro level of description via an intermediate meso level description is that we are able to *compute* the Markovianity measures $\langle I_1 \rangle$ and $\langle I_2 \rangle$ instead of performing an extensive series of numerical simulations. Namely, it is possible to compute the $\mu([k_{t-1}, k_t, k_{t+1}])$ and respectively the $\mu([k_{t-2}, k_{t-1}, k_t, k_{t+1}])$ on the basis of the meso chain $(\mathbf{X}, \tilde{P})$ which in turn is a loss-less description of the microscopic system (see Fig. 7.5).

Let us consider that for the cylinders of length 3. As noted above, the grammar $\mathfrak{G}_3$ of the system is determined by the fact that $|k_t - k_{t-1}| \leq 1$ and $|k_{t+1} - k_t| \leq 1$. Therefore, as illustrated in Fig. 7.9, for any $k_t = k$ with $0 < k < N$ there are nine possible paths $[k_{t-1}, k, k_{t+1}]$ and for $k = 0$ and $k = N$ there are respectively four paths. In order to compute the probability of a certain macro path, say $[p, k, f]$ p for past and f for future, we have to sum over all meso level paths that contribute to the given macro path. Let us denote a meso level path as $[(m_p l_p), (m l), (m_f l_f)]$ with $m_p + l_p = p, m \mid l = k$ and $m_f + l_f = f$. Its probability is given by

$$\mu([(m_p l_p), (m l), (m_f l_f)]) = \tilde{\pi}_{m_p, l_p} \tilde{P}(\tilde{X}_{m_p, l_p}, \tilde{X}_{m, l}) \tilde{P}(\tilde{X}_{m, l}, \tilde{X}_{m_f, l_f}). \tag{7.21}$$

The l.h.s. in Fig. 7.10 illustrates the possible paths for one $\tilde{X}_{m, l}$ with $m + l = k$. Notice that for a given macro state X_k there are $k + 1$ meso states if $k \leq M$ and respectively $N - k + 1$ meso states for $k > M$ (these numbers are for the case $M = L$ with $M + L = N$). In the case sequences of length three are considered, the situation is still quite clear. For instance, a macro path $[k - 1, k, k + 1]$ can be realized in four different ways for each[2] $\tilde{X}_{m, l}$ with $m + l = k$:

$$[(m - 1 \ l), (m \ l), (m + 1 \ l)] \tag{7.22}$$

$$[(m - 1 \ l), (m \ l), (m \ l + 1)]$$

$$[(m \ l - 1), (m \ l), (m + 1 \ l)]$$

$$[(m \ l - 1), (m \ l), (m \ l + 1)]$$

---

[2]Notice that the number of possibilities reduces at the corners or borders of the meso chain whenever $m = 0$ or $l = 0$.

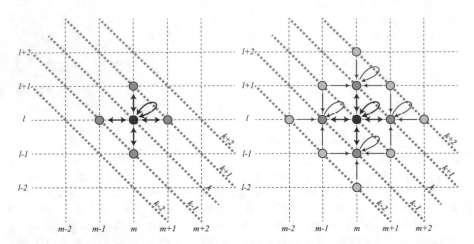

Fig. 7.10 Illustration of the possible paths for one $\tilde{X}_{m,l}$ with $m+l = k$ for cylinders of length three (*l.h.s.*) and four (*r.h.s.*) An *arrow* indicates whether or not one state can be followed by another in a sequence

The same reasoning can be applied to derive the probabilities for cylinders of length four even though the situation becomes slightly more complicated, as illustrated on the r.h.s. of Fig. 7.10.

Once the probabilities to for blocks of length three and four respectively, the computation of the Markovianity measures $\langle I_1 \rangle = \Delta H_2 - \Delta H_3$ and $\langle I_2 \rangle = \Delta H_2 - \Delta H_4$ is straightforward. All that is needed is to compute the respective block entropies.

7.4.2 Computation of Informational Flow

While the Markovianity measure informs us about the memory effects introduced at the macroscopic level, information flow measures the amount of information that knowledge about the micro level would add to the Markovian macroscopic formulation. In order to compute the information flow $I(X_{t+1} : \mathbf{x}_t | X_t)$ we have to compute $H(X_{t+1}|X_t) = \Delta H_2$ (see Eq. 7.3) as well as the conditional entropy $H(X_{t+1}|\mathbf{x}_t)$ between the micro and the macro level (see Eq. 7.4). For the two-community model, the micro states $\mathbf{x}$ are replaced by the respective meso-level state $\tilde{X}_{m,l}$ so that (7.4) reads

$$
\begin{aligned}
H(X_{t+1}|\tilde{X}_t) &= - \sum_{\tilde{X}_t \in \tilde{\mathbf{X}}} \tilde{\pi}_t \sum_{X_{t+1} \in \mathbf{X}} p(X_{t+1}|\tilde{X}_t) \log p(X_{t+1}|\tilde{X}_t) \\
&= - \sum_{\tilde{X}_{m,l} \in \tilde{\mathbf{X}}} \tilde{\pi}_{m,l} \sum_{k} Pr[X_f|\tilde{X}_{m,l}] \log Pr[X_f|\tilde{X}_{m,l}]
\end{aligned}
\tag{7.23}
$$

where the transition probabilities $Pr[X_f|\tilde{X}_{m,l}]$ are computed on the basis of the mesoscopic transition matrix $\tilde{P}$ as

$$Pr[X_f|\tilde{X}_{m,l}] = \sum_{m_f+l_f=f} \tilde{P}(\tilde{X}_{m,l}, \tilde{X}_{m_f,l_f}). \tag{7.24}$$

Notice that in analogy to the previous section and to make clear that it refers to transition probabilities[3] we use the subscript f to indicate the future state at time $t+1$.

The meso-to-macro probabilities $Pr[X_f|\tilde{X}_{m,l}]$ are illustrated in Fig. 7.11 which shows the transitions from the meso-level states with $m+l = k$ (X_k) to the neighboring macro set with $m_f+l_f = k+1$ (X_{k+1}). Notice that for any meso state $\tilde{X}_{m,l}$ there are, in general, only two possibilities to go to X_{k+1}: depending on

Fig. 7.11 Illustration of the meso-to-macro transitions $Pr[X_f|\tilde{X}_{m,l}]$ for two subsets X_k and X_{k+1} (so that $m+l = k$ and $m_f+l_f = k+1$) of the macroscopic partition. In the non-lumpable case the probabilities differ from one meso state to another

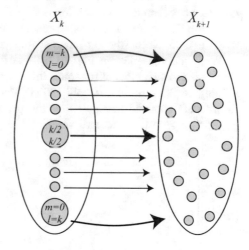

[3]This is to avoid a possible ambiguity because the probability $Pr[X_k|\tilde{X}_{m,l}]$ could also be read in terms of the projection from $\tilde{X}$ to X where $Pr[X_k|\tilde{X}_{m,l}]$ would indicate the probability with which the meso state $\tilde{X}_{m,l}$ is taken by ϕ to the macro state X_k.

the community in which an agent change happens it may transit to $\tilde{X}_{m+1,l}$ or $\tilde{X}_{m,l+1}$. An example for the transition from X_{50} to X_{51} has been shown in Sect. 7.3.2, Fig. 7.6. As the coupling across communities decreases (in relation to the intra-community coupling) the meso-to-macro transition probabilities become more heterogeneous. The reduction in uncertainty about the macroscopic evolution that information flow captures is due to this heterogeneity because $H(X_{t+1}|\tilde{X}_t)$ is small compared to the uniform lumpable case (i.e., for $r = 1$).

7.5   Results

We first look at the resulting information quantities as a function of the coupling between the two communities r for a system of $N = 100$ agents ($M = L = 50$). Figure 7.12 shows $\langle I_1 \rangle$ (dashed curves) and $\langle I_2 \rangle$ (solid curves) as a function of the coupling between the two communities r for a system of $N = 100$ agents ($M = L = 50$). The different curves represent various different contrarian rates p from 0.001 to 0.05. Notice the log-linear scaling of the figure.

What becomes clear in Fig. 7.12, first of all, is that the deviation from Markovianity is most significant for small inter-community couplings. This means, in the reading of Görnerup and Jacobi (2008), that the additional information about the future state (beyond that given by the present) provided by pasts of length n is larger than zero if r becomes small. In general and not surprisingly, $\langle I_2 \rangle > \langle I_1 \rangle$

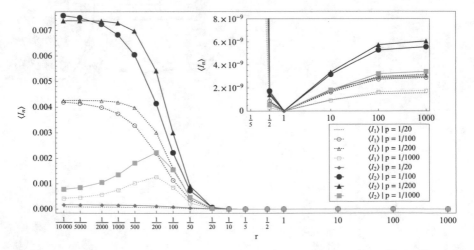

Fig. 7.12 $\langle I_1 \rangle$ (*dashed curves*) and $\langle I_2 \rangle$ (*solid curves*) as a function of the coupling between the two communities r for a system of $N = 100$ agents. The different *curves* represent various different contrarian rates p from 0.05 to 0.001, see legend

which means that both the first and the second outcome before the present provide a considerable amount of information. In fact, the numbers indicate that the first and the second step into the past contribute in almost the same way. Noteworthy, the two measures $\langle I_1 \rangle$ and $\langle I_2 \rangle$ behave in the same way from the qualitative point of view which suggests that the computationally less expensive $\langle I_1 \rangle$ can be well-suited for the general Markovianity test.

The inset in Fig. 7.12 shows the situation for values around $r = 1$ (homogeneous mixing) as well as $r > 1$. As we would expect by the strong lumpability of homogeneous mixing, $\langle I_1 \rangle$ and $\langle I_2 \rangle$ are effectively zero (order 10^{-17}) in the case $r = 1$. Also if the inter-community coupling becomes larger than the coupling within communities (a situation that resembles a bipartite graph) $\langle I_1 \rangle$ and $\langle I_2 \rangle$ are very small which indicates that a Markovian macro description (that is, ideal aggregation) describes well these situations.

Next, Fig. 7.13 compares the Markovianity measure $\langle I_2 \rangle = I(X_{t+1} : X_{t-1}, X_{t-2} | X_t)$ with informational closure $I(X_{t+1} : \tilde{X}_t | X_t)$. Here we consider two different contrarian rates $p = 0.001$ (green curves) and $p = 0.005$ (blue curves). We clearly observe that $I(X_{t+1} : X_{t-1}, X_{t-2} | X_t) < I(X_{t+1} : \tilde{X}_t | X_t)$ and remember that Pfante et al. (2014a) has proven that $I(X_{t+1} : X_{-\infty}^{t-1} | X_t) \leq I(X_{t+1} : \tilde{X}_t | X_t)$. It would of course be desirable to analyze the behavior of the Markovianity for longer histories in order to find out whether $I(X_{t+1} : X_{-\infty}^{t-1} | X_t)$ eventually converges to $I(X_{t+1} : \tilde{X}_t | X_t)$ or if their remains a gab between information flow from micro to macro and macroscopic memory.

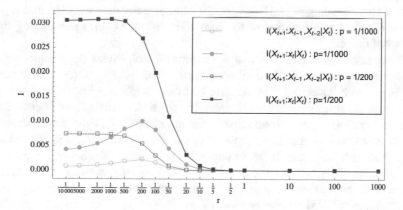

Fig. 7.13 Comparing the Markovianity measure $\langle I_2 \rangle = I(X_{t+1} : X_{t-1}, X_{t-2} | X_t)$ (*dashed*) with informational closure $I(X_{t+1} : \tilde{X}_t | X_t)$ (*solid*) as a function of the coupling between the two communities r for a system of $N = 100$ agents. The different *curves* represent two different contrarian rates $p = 0.001$ (*green*) and $p = 0.005$ (*blue*), see legend

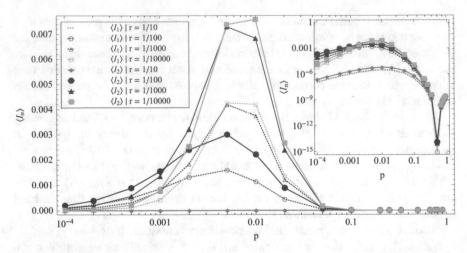

Fig. 7.14 $\langle I_1 \rangle$ and $\langle I_2 \rangle$ as a function of the contrarian rate p for various coupling ratios r and a system of $M = L = 50$

We notice in Figs. 7.12 and 7.13 that the measures do not generally increase monotonically with a decreasing ratio r which is most obvious for the example with a very small $p = 1/1000$ (green curves). This is somewhat unexpected and it indicates the existence of certain parameter constellations at which macroscopic complexity (for this is how non-Markovianity may be read) is maximized. To obtain a better understanding of this behavior, the Markovianity measures $\langle I_1 \rangle$ and $\langle I_2 \rangle$ are plot in Fig. 7.14 as a function of the contrarian rate p. Notice again the log-linear scaling of the plot.

It becomes clear that there is a strong and non-trivial dependence of the Markovianity measures on the contrarian rate p. Namely, $\langle I_1 \rangle$ and $\langle I_2 \rangle$ are very small if p is relatively large but they are also relatively small if p becomes very small. There is a parameter regime in between in which deviations from Markovianity become most significant. Notice that in the inset of Fig. 7.14 the same curves are shown on a double-logarithmic scale. This shows, first, that $\langle I_1 \rangle$ and $\langle I_2 \rangle$ for very small p are still significantly larger compared to the case of relatively large p (say $p > 0.1$). Secondly, we observe that $\langle I_1 \rangle$ and $\langle I_2 \rangle$ actually vanish for $p = 1/2$. As discussed in the previous section, the reason for that is the strong lumpability of the two-community CVM whenever $p = 1/2$.

Finally, a detailed picture of the dependence of $\langle I_n \rangle$ on the contrarian rate is provided in Fig. 7.15. The plot compares the cases $r = 1/100$ and $r = 1/1000$ in order to show that the peak in the $\langle I_n \rangle$ depend also on r. For the interpretation of this behavior, notice that the p at which deviations from Markovianity become largest, lie precisely in the parameter interval in which switching times between the two complete consensus states become minimal. Compare Fig. 6.4 in Sect. 6.3.3.

Fig. 7.15 Detailed picture of the dependence of $\langle I_n \rangle$ on the contrarian rate. *Blue curves* correspond to $r = 1/100$ and *red curves* to $r = 1/1000$. In the first case the peak is at around $p \approx 0.05$, in the latter at $p \approx 0.065$

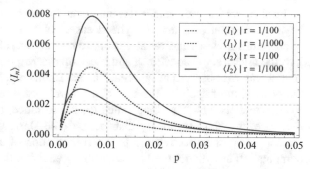

7.6 Summary and Discussion

This chapter has been devoted to the study of the non-Markovian case using the example of the two-community CVM. We have first considered the example from the lumpability point of view and examined the reasons for which strong and weak lumpability are not satisfied. Then, following the setting proposed in Pfante et al. (2014a) and Görnerup and Jacobi (2008), we have adopted an information-theoretic perspective in order to analyze the effects that an inhomogeneous interaction graph brings about at the macroscopic level.

Therefore, this analysis can be seen as a first step to measure the complexity introduced by aggregation without sensitivity to the micro- or mesoscopic structures. The two-community CVM, in which the population is composed of two sub-population of size L and M such that $L + M = N$ and it is assumed that individuals within the same sub-population are connected by strong ties whereas only weak ties connect individuals belonging to different communities, is a well-suited example to study aggregation artifacts for an inhomogeneous (though still very simple) interaction topology. While a lumpable description can be obtained by an independent aggregation of the two communities, leading to a Markov chain of $(\tilde{\mathbf{X}}, \tilde{P})$ manageable size $(O(N^2))$, aggregation over the entire agent population without sensitivity to the population structure leads to a description which is not lumpable. On the basis of the intermediate meso chain $(\tilde{\mathbf{X}}, \tilde{P})$ it is possible to explicitly compute the information flow from micro to macro $I(X_{t+1} : \mathbf{x}_t | X_t)$ (Pfante et al. 2014a) as well as the Markovianity measures $\langle I_1 \rangle = I(X_{t+1} : X_{t-1} | X_t)$ and $\langle I_2 \rangle = I(X_{t+1} : X_{t-1}, X_{t-2} | X_t)$ (Görnerup and Jacobi 2008) with respect to the non-lumpable macro description obtained by aggregation over the entire population. While information flow accounts for the information that a macroscopic formulation omits, the Markovianity measures show that this information is (at least in part) still present at the macro level in form of temporal correlations. This shows that information-theoretic measures are a promising tool to study the relationship between different levels of description in complex multilevel systems such as ABMs and that global aggregation over an agent population without sensitivity to micro- or mesoscopic structures leads to memory effects at the macroscopic level.

Further work is needed to better understand these results and extend the analysis beyond this relatively simple stylized case. By experiments with more complex interaction structures and variations of the interaction rules we may gain an insight into the microscopic conditions and mechanisms responsible for the temporal and spatial patterns observed at aggregate levels. In this regard, I would like to mention the possibility of applying the arguments developed throughout this chapter to the case of models with absorbing states as, for instance, the pure VM ($p = 0$). In that case, the quasi-stationary distribution (see Darroch and Seneta 1965) takes the role of $\hat{\pi}$ or respectively $\tilde{\pi}$ in the construction of an ideal aggregate and the computation of cylinder measures. Finally, I envision that the framework might be useful for the evaluation of different approximation schemes developed for complex high-dimensional agent models.

References

Ball, R. C., Diakonova, M., & Mackay, R. S. (2010). Quantifying emergence in terms of persistent mutual information. *Advances in Complex Systems, 13*(03), 327–338.

Banisch, S. (2014). From microscopic heterogeneity to macroscopic complexity in the contrarian voter model. *Advances in Complex Systems, 17*, 1450025.

Banisch, S., Lima, R., & Araújo, T. (2012). Agent based models and opinion dynamics as Markov chains. *Social Networks, 34*, 549–561.

Buchholz, P. (1994). Exact and ordinary lumpability in finite Markov chains. *Journal of Applied Probability, 31*(1), 59–75.

Burke, C. J., & Rosenblatt, M. (1958). A Markovian function of a Markov chain. *The Annals of Mathematical Statistics, 29*(4), 1112–1122.

Chazottes, J.-R., Floriani, E., & Lima, R. (1998). Relative entropy and identification of Gibbs measures in dynamical systems. *Journal of Statistical Physics, 90*(3–4), 697–725.

Chazottes, J.-R., & Ugalde, E. (2003). Projection of Markov measures may be Gibbsian. *Journal of Statistical Physics, 111*(5/6), 1245–1272.

Darroch, J.N., & Seneta, E. (1965). On quasi-stationary distributions in absorbing discrete-time finite Markov chains. *Journal of Applied Probability, 2*(1), 88–100.

Görnerup, O., & Jacobi, M. N. (2008). A method for inferring hierarchical dynamics in stochastic processes. *Advances in Complex Systems, 11*(1), 1–16.

Görnerup, O., & Jacobi, M. N. (2010). A method for finding aggregated representations of linear dynamical systems. *Advances in Complex Systems, 13*(02), 199–215.

Gurvits, L., & Ledoux, J. (2005). Markov property for a function of a Markov chain: A linear algebra approach. *Linear Algebra and its Applications, 404*(0), 85–117.

Jacobi, M. N., & Görnerup, O. (2009). A spectral method for aggregating variables in linear dynamical systems with application to cellular automata renormalization. *Advances in Complex Systems, 12*(02), 131–155.

James, R. G., Ellison, C. J., & Crutchfield, J. P. (2011). Anatomy of a bit: Information in a time series observation. *Chaos, 21*(3), 7109.

Kemeny, J. G., & Snell, J. L. (1976). *Finite Markov chains*. New York: Springer.

Ledoux, J., Rubino, G., & Sericola, B. (1994). Exact aggregation of absorbing Markov processes using the quasi-stationary distribution. *Journal of Applied Probability, 31*, 626–634.

Pfante, O., Bertschinger, N., Olbrich, E., Ay, N., & Jost, J. (2014a). Comparison between different methods of level identification. *Advances in Complex Systems, 17*, 1450007.

Pfante, O., Olbrich, E., Bertschinger, N., Ay, N., & Jost, J. (2014b). Closure measures for coarse-graining of the tent map. *Chaos: An Interdisciplinary Journal of Nonlinear Science, 24*(1), 013136.

Shalizi, C. R. (2001). *Causal architecture, complexity and self-organization in the time series and cellular automata.* (Doctoral dissertation, University of Wisconsin–Madison).

Shalizi, C. R., & Moore, C. (2003). What is a Macrostate? Subjective observations and objective dynamics. In *CoRR*. arXiv:cond-mat/0303625.

Vilela Mendes, R., Lima, R., & Araújo, T. (2002). A process-reconstruction analysis of market fluctuations. *International Journal of Theoretical and Applied Finance, 5*(08), 797–821.

Chapter 8
Overlapping Versus Non-overlapping Generations

In this chapter, we inspect well-known population genetics and social dynamics models. In these models, interacting individuals, while participating in a self-organizing process, give rise to the emergence of complex behaviors and patterns. While one main focus in population genetics is on the adaptive behavior of a population, social dynamics is more often concerned with the splitting of a connected array of individuals into a state of global polarization, that is, the emergence of speciation. Using numerical simulations and the mathematical tools developed in the previous chapters we show that the way the mechanisms of selection, interaction and replacement are constrained and combined in the modeling have an important bearing on both adaptation and the emergence of speciation. Differently (un)constraining the mechanism of individual replacement provides the conditions required for either speciation or adaptation, since these features appear as two opposing phenomena, not achieved by one and the same model. Even though natural selection, operating as an external, environmental mechanism, is neither necessary nor sufficient for the creation of speciation, our modeling exercises highlight the important role played by natural selection in the interplay of the evolutionary and the self-organization modeling methodologies.

8.1 Introduction

There are two important phenomena observed in evolutionary dynamical systems of any kind: *self-organization* and *emergence*. Both phenomena are the exclusive result of endogenous interactions of the individual elements of an evolutionary dynamical system. Emergence characterizes the patterns that are situated at a higher macro level and that arise from interactions taking place at the lower micro level of the system. Self-organization, besides departing from the individual micro interactions, implies an increase in order of the system, being usually associated to the promotion

© Springer International Publishing Switzerland 2016
S. Banisch, *Markov Chain Aggregation for Agent-Based Models*,
Understanding Complex Systems, DOI 10.1007/978-3-319-24877-6_8

of a specific functionality and to the generation of patterns. Typically, complex patterns emerge in a system of interacting individuals that participate in a self-organizing process. Self-organization is more frequently related to the process itself, while emergence is usually associated to an outcome of the process.

Although less frequently mentioned, the emergence of patterns from self-organizing processes may be strongly dependent on *locality*. Emergence and self-organization are not enough to distinguish between two important and quite different circumstances: the presence of an influence that impacts the system globally and, conversely, the absence of any global influence and the lack of information about any global property of the system. In the latter case, the system itself is the exclusive result of local interactions.

Such a global influence (entity or property) is often associated with the concept of *environment*. Noteworthy, the latter circumstance may be considered a case of the former: when no global entity exists, the environment for each agent is just the set of all the other agents. Conversely, when a global entity exists, it is considered part of the environment and may have an inhomogeneous impact on the individual dynamics.

Regardless of the environmental type, economical, ecological and social environments share as a common feature the fact that the agents operating in these environments usually try to improve some kind of utility, related either to profit, to food, to reproduction or to comfort and power. A general concept that is attached to this improvement attempt is the idea of *adaptation*.

In the economy, adaptation may be concerned with the development of new products to capture a higher market share or with the improvement of the production processes to increase profits: that is, innovation. In ecology, adaptation concerns better ways to achieve security or food intake or reproduction chance and, in the social context, some of the above economical and biological drives plus a few other less survival-oriented needs. In all cases, adaptation aims at finding strategies to better deal with the surrounding environment (Araújo and Vilela Mendes 2009).

Natural selection through fitness landscapes or geographic barriers are good examples how global influences are considered when modeling adaptation in an evolutionary process. On the other hand, adaptation also operates in many structure generating mechanisms that can be found in both physical and social sciences but that are built on the exclusive occurrence of local interactions.

In biology, the ultimate domain of evolution and natural selection, we are confronted with tremendous organic diversity—virtually infinite forms and shapes none of which found twice—but the distribution is well-structured in a way that allows us to order this diversity and to speak of species, families, orders etc. A quite illustrative description is given by the evolutionary geneticist Theodosius Dobzhanski (1970, p. 21):

> Suppose that we make a fairly large collection, say some 10,000 specimens, of birds or butterflies or flowering plants in a small territory, perhaps $100\,km^2$. No two individuals will be exactly alike. Let us, however, consider the entire collection. The variations that we find in size, in color, or in other traits among our specimens do not form continuous distributions. Instead, arrays of discrete distributions are found. The distributions are separated by gaps,

that is, by the absence of specimens with intermediate characteristics. We soon learn to distinguish the arrays of specimens to which the vernacular names English sparrow, chickadee, blue jay, blackbird, cardinal, and the like, are applied.

If we had to make a visual representation of this description of intra- and inter-species variations it would perhaps look like the multi-modal distribution shown in Fig. 8.1. What we call a species, is in fact some norm or mean characteristics of a cluster of individuals.

Evolutionary theory is ultimately a theory about the history which led to such a pattern. And if the organic diversity we observe nowadays evolved in a way that is characterized by some kind of "Tree of Live", then there must be events that may lead to the split of a connected set of individuals (protospecies) into (at least) two sets that are not connected any longer (see Fig. 8.2). In biology, this is called *speciation*. As we will see in this chapter, though, the generation of such a split with simple but well-known evolutionary models in which "natural selection impels and directs evolutionary changes" (Dobzhanski 1970, p. 2) is not straightforward. It so happens that constraints on the interaction behavior are required.

The phenotype of living beings is not the only domain where patterns of structured diversity as illustrated in Fig. 8.1 are observed. Phenomena include

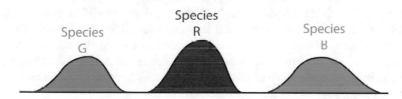

Fig. 8.1 Schematic illustration of organic diversity

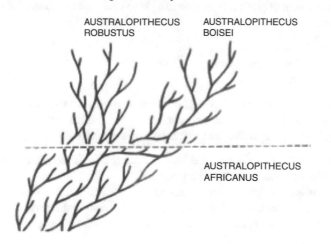

Fig. 8.2 Illustration of a speciation event. I am are grateful to Andreas Dress for providing me this figure

certain phases of structure formation in physical cosmology, distribution of cultural behavior, languages and dialects, herd behavior in finance, among others.

Especially for the latter examples in the field of socio-cultural dynamics a variety of models has been proposed which do not rely on the evolutionary concept of (natural) selection.[1]

They are rather based on the idea of exclusively *Local Interactions (LI)* implemented in form of a system of agents that interact locally according to simple rules like assimilation or conformity. In these systems, finding strategies to better deal with the surrounding environment (and thus improving fitness) is not constrained by any global property. It may, however, be constrained by local (individual) rules.

As we shall see later in this chapter, constraints on the mechanisms of selection, interaction and replacement and the way they are combined in the modeling of an evolutionary process have an important bearing on both adaptation and emergence of speciation. Locality operating in each of these mechanisms seems to be the fundamental modeling principle by which emergence of a multi-modal distribution as shown in Fig. 8.1 can be explained. On the basis of these observations about the "modelability" of speciation with evolutionary and self-organisatory models, we study in this chapter the conditions and mechanisms required for speciation and the emergence of a multi-modal distribution.

In this analysis, we rely on computational models (Sect. 8.2) and apply the mathematical tools developed throughout this book to simplified versions of the models to understand these numerical findings (Sect. 8.3). Our models simulate how a population of individuals evolves in time in an abstract attribute space (S) that represent phenetic traits, attitudes, verbal behavior, etcetera. Modeling agents as points in an attribute space of this kind is of course a highly artificial abstraction from the complexity and multi-dimensionality of real agents.

For the purposes of this chapter, let us conceptualize an *interaction event*, defining the system evolution from one time step to the other, by the following three components:

1. selection of agents,
2. application of interaction rules,
3. replacement of agents.

Any interaction event (e.g., mating, communication,...) that takes place in the course of a simulation of the model consists of the sequential application of these three steps. The reason to dissect the interaction events in this way is twofold:

1. we want to look at the dynamical and structural effects of constraints applied to each of the three components independently;
2. the scheduling of interaction events may have a crucial effect on the model behavior, and with the distinction between selection and interaction on the one hand, and replacement on the other, we are able to make this effect explicit.

---

[1]See Castellano et al. (2009) for a comprehensive overview over models in this field.

The way interaction events are scheduled in the implementation of the models is not always given much importance in existing simulation studies. In the presence of constraints on the selection and interaction mechanisms, however, the outcome as well as the dynamical properties depend in a crucial way on the different choices. On the other hand, there are studies that do analyze the differences between synchronous and asynchronous update (see, for instance, Huberman and Glance 1993; Banisch 2010) as well as studies on non-overlapping (NOLG) and respectively overlapping generations (OLG) in biology and economics (for instance, Kehoe and Levine 1984).

Here we show that especially when the interaction is constrained (as in the case of assortative mating) there emerges an important qualitative difference between OLG and NOLG models. Namely, *speciation is observed in the former, but not in the latter case, whereas adaptation is favored by the latter and hindered by the former.* However, by the distinction of selection, interaction and replacement we are able to show that in fact the difference between local and non-local replacement plays the determinant role (and not the distinction between OLG and NOLG). Even though locality also impacts selection and interaction mechanisms, it is on the replacement mode where relies the fundamental difference with respect to the conditions required for either adaptiveness or speciation.

The chapter is organized as follows: Sect. 8.2 addresses the main issues of both the fitness landscape and the self-organizing models from a computer simulation framework. In both cases, microscopic implementation rules are tested against their capability of reproducing adaptiveness and speciation. In Sect. 8.3, the emergence of speciation is analytically shown to be dependent on the choice of different replacement modes. This is accomplished through a probabilistic description of a minimal model of just three phenetic traits where the transition probabilities between traits follow a Markov chain. Section 8.4 is targeted at presenting concluding remarks and a framework that relates interaction events to the emergence of collective structures in adaptive and self-organizing complex systems.

8.2   From Adaptive Dynamics to Cluster Formation

8.2.1   Adaptive Walks on Fitness Landscapes

In biology, and population genetics in particular, adaptive walks on fitness landscapes have been studied intensively. The main questions addressed by fitness landscapes approaches are related to the possible structure of the landscapes (e.g., Kauffman 1993), to how populations climb an adaptive peak in the landscape (e.g., Fisher 1930), and to the circumstances under which a population might wander from one peak to another by crossing adaptive valleys (e.g. Wright 1932).

One of the best-known models for populations on fitness landscapes is the Wright-Fisher model with non-overlapping generations (sometimes called

Wright-Fisher sampling and shortened in the sequel by WF model, see Crow and Kimura 1970 and also Drossel 2001). Noteworthy, the WF model is an important predecessor of the voter model (VM), which has been studied most extensively throughout this work. Consider a population of N individuals which is said to constitute the original generation ($g = 0$). We consider only the case of sexual reproduction here, in which the genotype of a new-born individual is obtained by the recombination of the genomes of two randomly chosen parent individuals. As noted above, the choice of two parents and the application of a recombination rule is referred to as interaction (or mating) event. In the WF model, N such mating events are performed until a new generation of N individuals is complete. As soon as it is complete, the parent generation is canceled and the process is repeated taking the new generation as parents. Therefore, in the WF model the population size is always maintained at N. We will denote the generation number by $g = 0, 1, 2, \ldots$.

We implemented this simple model and performed simulations on different toy fitness landscapes. The microscopic rules involved into the creation of a new individual, that is, the mating event, are as follows:

1. selection of two individuals with a probability proportional to their fitness,
2. application of recombination and mutation rules,
3. replacement of an agent from the parent generation.

In this toy model, we consider only one phenetic trait (locus) that takes discrete values (from 0 to 99). We denote the traits of the two chosen parent individuals i and j as x_i and x_j respectively and model recombination by taking the average of the two, $x_{new} = (x_i + x_j)/2$. To model mutations we add a random value to x_{new}. In the WF model, x_{new} is stored at an arbitrary place in the children array and this chapter will clarify that this has important consequences for the model dynamics.

An adaptive landscape is introduced into the model by assigning a fitness value to each of the 100 traits. For the first analysis shown in Fig. 8.3, a single-peaked

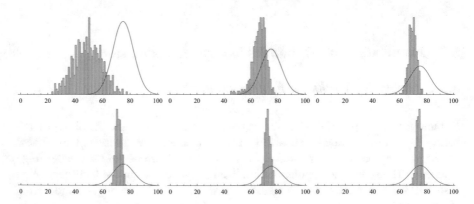

Fig. 8.3 WF model approaches an adaptive peak. In the *upper row* the initial generation ($g = 0$) and the distribution after the first ($g = 1$) and the second ($g = 2$) iteration are shown from *left to right*. *Bottom row* shows, from *left to right*, the 5th, 10th and 20th generation ($g = 5, 10, 20$)

fitness function with a peak at trait 75 is used and the fitness assigned to trait x is given by

$$F(x) = \frac{1}{15} e^{-\frac{2}{225}(-75+x)^2} \sqrt{\frac{2}{\pi}} = N(\mu, \sigma^2). \tag{8.1}$$

We have used the normal distribution with $\mu = 75$ and $\sigma^2 = 7.5$ in the construction of the fitness landscape (solid line in Fig. 8.3). In the iteration process, individuals are chosen as parents with a probability proportional to $F(x)$, x being the trait of the respective individual.

For the illustrative model realizations in this section, we set $N = 500$. Initially, the 500 individuals are distributed in this space according to a normal distribution with mean $\mu = 50$ and $\sigma^2 = 10$ (see first image of Fig. 8.3).

This section is mainly thought as an illustration of the different behaviors and patterns generated by certain constraints on the interaction mechanism. As the qualitative effects of different assumptions become evident and comprehensible in single simulations of the model, there is no need for a rigorous statistical analysis of suites of simulations with varying initial conditions. Moreover, a mathematical analysis of the model dynamics is presented in the second part of this chapter (Sect. 8.3).

Figure 8.3 shows the first few iterations of the WF model. The adaptive peak at around 75 is reached within only a few iterations. Due to mutations, the population does not become fixed at one specific trait, but maintains a certain amount of variation. Populations simulated with the WF model are very fast in reaching an adaptive peak in the fitness landscape.

8.2.2 Sympatric Speciation

Figure 8.3 shows that the WF model is well-suited to show how a finite population approaches a peak in the fitness landscape. However, what about speciation? To get a first insight about whether the splitting of the unimodal initial distribution into a bimodal distribution with two clusters is possible we simulated the model with a two-peaked fitness landscape. So the difference with respect to the previous simulation is that the fitness function (solid line) has two adaptive peaks, one centered at 25 and the other at 75. The fitness (that is, the probability of choosing an individual in state x) is defined by a mixture of two normal distributions $N(25, 7.5)$ and $N(75, 7.5)$:

$$F(x) = \frac{e^{-\frac{2}{225}(-75+x)^2}}{15\sqrt{2\pi}} + \frac{e^{-\frac{2}{225}(-25+x)^2}}{15\sqrt{2\pi}}. \tag{8.2}$$

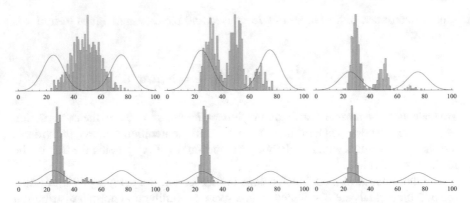

Fig. 8.4 WF model with a two-peaked fitness landscape approaches a single adaptive peak. From *top left to bottom right* the initial state ($g = 0$) and the first five generations are shown ($g = 1, 2, 3, 4, 5$)

The first five iterations of that model are shown in Fig. 8.4. The initial distribution is as in the previous example. We see that multi-modal shapes emerge only in the very first few generations of the model. Namely, after the first and the second iteration, there are three clusters: two located at the peaks and a third one with low fitness in between the other two. The latter can be seen as hybrid individuals with strong selective disadvantages that are obtained by a recombination of individuals from the different peaks. However, the disappearance of clustering is very fast and after only four iterations all the population concentrates at one of the peaks. Hence, in the model it is difficult to generate a stable co-existence of species.

The case considered here is the case of speciation in sympatry: no geographic constraints are assumed to divide the population into reproductive islands or to constrain the mating chances of pairs of individuals in any other way. A possible explanation why the simulation of sympatric speciation is not possible in the WF model as described above is provided in the seminal paper on sympatric speciation by Smith (1966). Smith showed that besides selective forces, it is necessary that the population sizes of the (two) sub-populations are regulated independently. Because the total population size is usually constant in the WF model (in our case $N = 500$), it does not implement an independent regulation of sub-populations.

An issue frequently discussed in the context of sympatric speciation is assortative mating (see, for instance, Kondrashov and Shpak 1998; Dieckmann and Doebeli 1999 and references therein). We also simulated the WF model with the additional constraint that two individuals need to be similar in order to produce offspring. Two chosen individuals i and j only produce an individual for the new generation if the their difference is small (here $|x_i - x_j| < 10$). The microscopic rules become:

1. selection of two individuals with a probability proportional to their fitness,
2. application of recombination and mutation rules *if the individuals are similar*,
3. replacement of an agent from the parent generation.

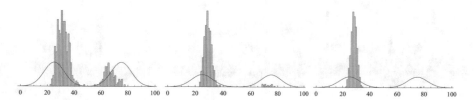

Fig. 8.5 WF model with a two-peaked fitness landscape and assortative mating approaches a single adaptive peak. From *left to right* the first three generations are shown ($g = 1, 2, 3$). The initial state is as in Fig. 8.4

In Fig. 8.5 the first three generations obtained by the iteration of this model are shown. The only difference with respect to the pure random mating case (Fig. 8.4) is that the intermediate cluster does not appear because the interbreeding of a pair of individuals from either peak is prohibited by the assortativity condition.

8.2.3   Cluster Formation in Opinion Dynamics

From the point of view of self-organizing systems in opinion or cultural dynamics (e.g. Axelrod 1997; Deffuant et al. 2001) the result shown in Fig. 8.5 is somewhat surprising because the introduction of interaction constraints is known to lead to co-existence of clusters of individuals (assortative mating is often called bounded confidence in this context). This is even more interesting as the microscopic rules used to model the self-organization in opinion dynamics are very similar.

1. selection of two individuals, *all with equal probability.*
2. application of recombination and mutation rules *if the individuals are similar,*
3. *update* of one parent agent.

In this scheme, we emphasized differences with respect to the WF model. Notice that mutations, sometimes interpreted as cultural drift, are not always taken into account.

Notice also that this form of replacement where effectively one parent individual is chosen to die to make place for the new-born is sometimes considered in population genetics (see, for instance, Moran 1958; Korolev et al. 2010).

In opinion dynamics the initial population is usually distributed according to the uniform distribution. In general, there are no global influences such as a fitness landscape so that the probability of selection is equal for all individuals independent of their position in the trait space.

The locally-interacting model (henceforward called LI model) is implemented as a model of overlapping generations (OLG). That is, the population is updated after each single interaction event (and not after N events). Notice that this means that the new state of an individual that is updated is from then on taken into account in the later iterations. Therefore, a single iteration actually means a single interaction

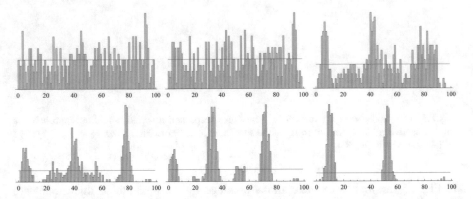

Fig. 8.6 Emergence of clustering in the LI model with assortativity. In the *upper row* the initial population and the distribution of the first and the tenth generation are shown from *left to right* ($g = 0, 1, 10$). *Bottom row* shows, from *left to right*, the 20th, 40th and 100th generation ($g = 20, 40, 100$)

event involving two individuals. Nevertheless, for the sake of comparability with the WF model, we can consider generations in the LI model by assuming that we pass from one generation to the next ($g \rightarrow g + 1$) after N iterations (interaction events).

In Fig. 8.6 we show a realization of the simulation for 500 individuals initially distributed uniformly over the traits from 0 to 99. Update only takes place if the distance between two individuals is smaller then 10. It becomes clear that initial inhomogeneities are reinforced during the process such that clusters of individuals are formed. Compared to the WF simulations this process is slow. In Fig. 8.6 we show from top left to bottom right the original population ($g = 0$), and the population in the 1st, 10th, 20th, 40th and 100th generation ($g = 1, 10, 20, 40, 100$). From generations 40–100 some of the clusters have disappeared so that only two large sub-populations (and a very small one at around 90) remain. In the long run these clusters might merge due to mutations (drift). In any case the co-existence of "reproductively isolated" sub-populations is rather stable during long periods of the process.

8.2.4 Overlapping Versus Non-overlapping Generations

There seems to be a subtle difference between the LI model and the WF model, with a crucial effect, however. There are three potential sources of the different behavior:

1. There is uniform fitness in the LI model but a peaked landscape in the WF model.
2. The LI model is implemented as a model of OLG whereas the WF model implements NOLG.
3. In the LI model, the state change of an individual is modeled whereas the creation of a new individual is considered in the WF model.

Fig. 8.7 WF model with overlapping generations with flat fitness landscape, assortative mating and uniform initial population as in Fig. 8.6. From *left to right* the first three generations ($g = 1, 2, 3$) are shown

The first two cases can be checked easily by implementing the WF model with OLG and looking at a realization using the same conditions as in Fig. 8.6. This is sometimes referred to as Moran model (Moran 1958; Korolev et al. 2010). Two individuals give rise to a new individual which replaces another individual in the current generation. Notice that the assortativity constrained (interaction only takes place if $|x_i - x_j| < 10$) is included. The first three generations ($g = 1, 2, 3$ corresponding to the population after 500, 1000 and 1500 sequential mating events) are shown in Fig. 8.7. The initial population is the same as before (upper left in Fig. 8.6). The behavior of the model is in drastic contrast to the behavior of the LI scheme. In fact, the behavior is very similar to the original WF model with NOLG.

We conclude that the qualitative differences between the WF model and the LI model are neither due to different ways of dealing with generations (OLG versus NOLG) nor to the choices of different fitness landscapes.

8.2.5   Local Versus Non-local Replacement

It turns out that in the implementation of the WF model with OLG, a decision must be taken whether the new individual replaces one of its parents or an arbitrary individual from the generation and that the two alternatives result in qualitatively different dynamical behaviors. We will call *local replacement* the case that the new individual replaces one of its parents and *non-local replacement* refers to the case that an arbitrary individual is replaced by the new one. Noteworthy, there is a tendency that models with NOLG implement a form of non-local replacement because no care is usually taken about the order of individuals such that a child will in general appear at a position in the population array that is distant from the position of the parents.

In this way, non-local replacement undermines the effects of assortative mating, because an individual with a trait x_i can effectively be replaced by an individual with trait x_j even if $|x_i - x_j| > 10$.

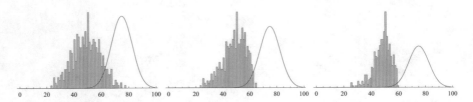

Fig. 8.8 The self-organization LI model with local update on a peaked fitness landscape. From *left to right* the initial population and the first two generations ($g = 0, 1, 2$) are shown

8.2.6 (Non-)adaptiveness of Local Replacement

Earlier we saw that modeling speciation in a model with a fixed population size requires that the update process operates with local replacement. However, it turns out that in this case the process looses its adaptiveness. Figure 8.8 shows the first generations ($g = 0, 1, 2$) of the self-organization model with local replacement performed on a fitness landscape with a single peak (compare Fig. 8.3). The population is actually pushed away from the peak. This is due to the fact that the individuals close to the peak, though frequently chosen, are not replaced by individuals with low fitness (rarely chosen) so that the proportion of fit individuals does not increase. To the contrary, mutations tend to drive fittest individuals away from the peak. Hence, the mode of replacement in these two models with almost the same microscopic rules has a dramatic effect on the dynamics behavior. Cluster formation (or speciation) and adaptiveness are in the context of these models two opposing phenomena such that an explanation of the two together is not achieved by one and the same model.

8.3 Probabilistic Analysis of a Minimal Model

The simulations show that there are decisive differences between different implementations of the simulation models even though the microscopic rules of agent choice and recombination are in fact equal. In particular, it turned out that the qualitative differences in the model behavior are due to different modes of agent replacement. This section elaborates these differences for a minimal model where the number of allowed traits is reduced from 100 to three. The model implements the same mechanisms as before, on this reduced space with three traits only, excepting mutations. Looking at the rate (probabilities) of transitions from one trait to the other we derive Markov chains, and the transition structure of these chains inform us about the dynamic mechanisms which different replacement modes give rise to.

8.3.1 A Minimal Model

Consider that there are only three different phenetic traits: the states left (L), right (R) and intermediate (M). As before, in every interaction event pairs of individuals are chosen and the state of the new individual is determined by the recombination of the parent states. We do not consider mutations here. In accordance with the recombination rule in the previous section, whenever two parents are in the same state, the child will also be in that state: we denote this by $LL \rightarrow L, MM \rightarrow M$ and $RR \rightarrow R$. If one of the parents is in L and the other in R recombination will lead to M, that is, $LR \rightarrow M$ and $RL \rightarrow M$. In case L mates with M we say that recombination leads to M $(LM \rightarrow M)$ and vice versa if M mates with L it leads to L $(ML \rightarrow L)$. Likewise, for matings between M and R-agents, we set $RM \rightarrow M$ and vice versa $MR \rightarrow R$. Notice that in the case of complete, homogeneous mixing the choice probabilities are symmetric such that choosing two agents with RM is equally likely as choosing them in reverse order MR. Notice that this model is very similar to the three-state VM studied in Chap. 4.

Associated with each of these nine possible transitions we define an additional probability α to be the probability that the recombination step is indeed performed once the respective trait combination is chosen. In the model without trait-dependent mating constraints or fitness differences, all the α are set to one. The reason for introducing this probability is that we can model assortative mating by setting $\alpha_{LR} = \alpha_{RL} = 0$. In that case, a pair of individuals in L and R are assumed to be unable to produce offspring. Because no state changes take place in that case, the respective probabilities now contribute to keeping whole population unchanged.

8.3.2 Transition Rates

We consider a system of N agents and characterize a population by counting the number of agents in the respective states L, M and R. Let us denote the number of agents in state L by l, the number of M-agents by m and the number of agent in R by r. After a mating event, the counters l, m, r are either unchanged or one of them increases while another one decreases by one (ex. $l-1, m+1$). The latter case simply means that one new individual (in state M) has replaced another one (in state L). In the case of complete mixing, all the agents have equal probabilities to be chosen in the iteration process. Therefore, we model the choice of an individual as a choice from an urn with N balls of three different colors L, M and R. It is then clear that the choice of an individual with feature F is f/N (generic F).

In this way, it is possible to derive equations for the probabilities of all the possible changes of l, m and r from one mating event to the other. Notice that already Moran adopted a similar Markov chain approach in the analysis of his model in Moran (1958). A careful consideration of the relation between these macro-level equations and the microscopic simulation model has been at the heart of the preceding chapters.

Let us denote the probability that m increases while l decreases by one as P_l^m. For the model with local replacement we use the convention that the agent chosen first is always replaced by the new one. Under this assumption the event $(l, m) \rightarrow (l-1, m+1)$ takes place if the states of the agent pair are either (L, R) or (L, M). The probability that a pair (L, R) is chosen is $\frac{lr}{N^2}$ which we denote as p_{LR}.[2] For (L, M) we have $p_{LM} = \frac{lm}{N^2}$. We integrate into this description the additional constraint α_{LR} (α_{LM}) as the probability that the respective combination, once chosen, gives indeed rise to a new individual. Then we obtain for probability $Pr[LR \rightarrow M] = \alpha_{LR}\, p_{LR}$ ($Pr[LM \rightarrow M] = \alpha_{LM}\, p_{LM}$). With this definitions we obtain

$$P_l^m = \alpha_{LR}\frac{lr}{N^2} + \alpha_{LM}\frac{lm}{N^2} = \alpha_{LR}\, p_{LR} + \alpha_{LM}\, p_{LM}. \tag{8.3}$$

Equivalently, for the other non-zero transitions in the local case we find

$$\begin{aligned}
P_r^m &= \alpha_{RL}\, p_{RL} + \alpha_{RM}\, p_{RM} \\
P_m^l &= \alpha_{ML}\, p_{ML} \\
P_m^r &= \alpha_{MR}\, p_{MR}.
\end{aligned} \tag{8.4}$$

For the model with non-local (random) replacement we assume that the new-born individual replaces a randomly chosen agent. The probability that this is an agent in state F is again f/N (generic F). With this convention we find for the replacement of an L-agent

$$\begin{aligned}
P_l^m &= \tfrac{l}{N}(\alpha_{LR}\, p_{LR} + \alpha_{RL}\, p_{RL}+ \\
&\quad + \alpha_{LM}\, p_{LM} + \alpha_{RM}\, p_{RM} + \alpha_{MM}\, p_{MM}) \\
P_l^r &= \tfrac{l}{N}(\alpha_{RR}\, p_{RR} + \alpha_{MR}\, p_{MR}).
\end{aligned} \tag{8.5}$$

For the replacement of an R-agent we have

$$\begin{aligned}
P_r^m &= \tfrac{r}{N}(\alpha_{LR}\, p_{LR} + \alpha_{RL}\, p_{RL}+ \\
&\quad + \alpha_{LM}\, p_{LM} + \alpha_{RM}\, p_{RM} + \alpha_{MM}\, p_{MM}) \\
P_r^l &= \tfrac{r}{N}(\alpha_{LL}\, p_{LL} + \alpha_{ML}\, p_{ML}),
\end{aligned} \tag{8.6}$$

---

[2]Notice that the agent choice is with replacement so that an individual may be chosen twice. This corresponds to self-fertilization and we allow it to keep the model as simple as possible.

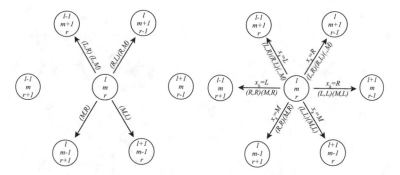

Fig. 8.9 Possible transitions in terms of the counters l, m, r for local (*l.h.s.*) and random (*r.h.s.*) replacement

and for replacement of an M-agent

$$P^l_m = \frac{m}{N} \left(\alpha_{LL} \, p_{LL} + \alpha_{ML} \, p_{ML} \right)$$
$$P^r_m = \frac{m}{N} \left(\alpha_{RR} \, p_{RR} + \alpha_{MR} \, p_{MR} \right).$$

$$(8.7)$$

For a better orientation we visualize the possible transitions for both replacement modes along with the conditions for the transitions in Fig. 8.9.

8.3.3   Random Mating

If all the α are equal to one, the transition equations (8.3) and (8.4) and respectively Eqs. (8.5)–(8.7) realize all the transitions shown in Fig. 8.9 with a probability greater than zero. In Fig. 8.10 the complete transition structure is shown for the model with five agents. Notice that for $N = 5$ each counter (l, r, m) can take values in between zero and five and that the triangular structure appears because we have $l+m+r = N$.

The larger gray atoms are the absorbing states of the process: they can be reached by a transition, but once reached, there is no transition leaving them. Therefore they characterize the final configurations of the process. For both local and non-local replacement the absorbing states are the three corners of the triangle grid with $l = N$ or $m = N$ or $r = N$. This means that the process will converge to a population with all individuals in the same state. The smaller light-blue states indicate the transient atoms and the chances that the process remains in those atoms decreases exponentially with time (see, for instance, Seneta 2006, and Chaps. 4 and 5).

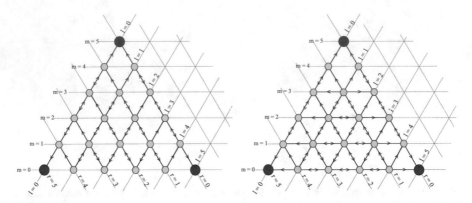

Fig. 8.10 Transition structure for local replacement (*l.h.s.*) and random replacement (*r.h.s.*) with random mating. In both cases, there are three absorbing states each corresponding to a homogeneous population

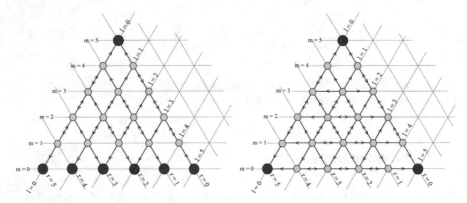

Fig. 8.11 Transition structure for local replacement (*l.h.s.*) and random replacement (*r.h.s.*) with assortative mating. Additional absorbing states emerge under local replacement, but not under random replacement as used in the WF model

8.3.4 Assortative Mating

The situation becomes different if we set $\alpha_{LR} = \alpha_{RL} = 0$ by which we prohibit mating between L- and R-agents. This is assortative mating which means, in this simple model with only three traits, that left and right agents are incompatible and cannot produce offspring. As noted above, the respective probabilities contribute to the probability that nothing changes as in that cases $(l, m, r) \rightarrow (l, m, r)$. For both replacement modes the assortativity condition changes the transition probabilities and we compare the resulting transition structures in Fig. 8.11.

Most importantly, for the local model, all the probabilities in (8.3) and (8.4) become zero if $m = 0$. Hence, if there is no intermediate individual left ($m = 0$) the process will remain where it is even if both l and r are larger than zero. Assortative mating may therefore lead to the stable co-existence of L- and R-agents (just as it has been observed for the three-state VM, see Chap. 4 and Banisch et al. 2012). Under non-local replacement this does not happen because even if certain transitions are canceled there remain horizontal transitions leading away from the respective two-species configurations. This explains why speciation cannot be observed in the simulations performed in the first part of this chapter.

It so happens that random replacement sets aside the effects of bounded confidence and consequently—like in the case of undirected genetic drift—leads to the merging of subpopulations. As random interbreeding contributes to conservative dynamics, random replacement is also an opposing force to speciation. This is due to the fact that under this replacement mode, a newcomer agent may take the place of a former-distant one. In so doing, forbidden transitions turn out to be allowed so that the consequences at the macro level become the same of unbounded confidence.

8.3.5 Two-Peaked Fitness Landscape

Next, let us discuss an extreme case of a two-peaked fitness function. We consider the case that intermediate individuals have a zero fitness which we model by prohibiting all matings in which M-agents are involved. This situation can be obtained by assigning a zero probability to all the respective transitions, that is:
$\alpha_{LM} = \alpha_{ML} - \alpha_{MM} = \alpha_{RM} = \alpha_{MR} = 0$

From Eq. (8.4) we see that in the case of local replacement this leads to the strange situation that the probabilities for all those transitions by which the number of intermediates decreases become zero: $P_m^l = P_m^r = 0$. Unless initialized with all agents in L or R, the simulation will converge to the situation where all individuals are in the intermediate state (M), with zero fitness. This clearly points at a deficiency of modeling adaptive dynamics with local replacement.

All in all, we can conclude that adaptiveness is favored by non-local replacement while it is difficult to achieve speciation. As opposed to this, under local replacement speciation becomes a natural result of assortative mating, but then the process is not convenient for approaching adaptive peaks in a fitness landscape.

8.4 Summary and Discussion

In the context of the models we studied in this chapter, evolutionary models that build on natural selection and locally interacting dynamics building on self-organization do not appear as opposing one another. In fact, the dynamical update rules used in the modeling of the microscopic interactions follow the same principles.

Let us try to adopt a broader perspective and to figure out a general framework comprising the main mechanisms leading to the emergence of collective structures

Table 8.1 Summary of the effects that constraints imposed on the selection, interaction and replacement step lead to

Mechanisms			Emergent patterns	
Selection	Interaction	Replacement	Outcome	Example
1 peak	Random	Random	Convergence with adaptation	Figure 8.3
2 peaks	Random	Random	Convergence with adaptation	Figure 8.4
1 peak	Assortative	Random	Convergence with adaptation	Figure 8.5
Random	Assortative	Local	Speciation	Figures 8.6 and 8.11(left)
Random	Assortative	Random	Convergence	Figures 8.7 and 8.11(right)
1 peak	Assortative	Local	Convergence without adaptation	Figure 8.8
2 peaks	Assortative	Local	Convergence without adaptation	—(Sect. 8.3.5)
Random	Random	Local	Convergence	Figure 8.10(left)
Random	Random	Random	Convergence	Figure 8.10(right)

in adaptive and self-organizing complex systems. Back to the two phenomena that we address in this chapter, we may say that the main consequences to the "modelability" of either adaptation or speciation are due to the constraints imposed on each of the mechanisms of selection, interaction and replacement. Their interplay is summarized in Table 8.1.

The framework presented in Table 8.1 schematically shows the consequences of adopting (un)constrained mechanisms to the emergent outcome of a self-organizing process. It helps to emphasize that the emergence of some specific patterns may be strongly dependent on the way constraints dictate limitations on the selection, interaction and replacement mechanisms. More specifically, it shows that differently (un)constraining the replacement mechanism of a process provides the conditions required for either speciation (the emergence of multi-modal distributions) or adaptation, since these features appear as two opposing phenomena, not achieved by one and the same model.

In the same way that random interbreeding leads to conservative dynamics, random replacement is also an opposing force to speciation since newcomers may take the place of formerly distant agents. At the macro level, random replacement sets aside the effect of bounded confidence and—like undirected genetic drift—may lead to the merging of subpopulations. This sheds light on the fact that some of the differences observed between models with overlapping versus models with non-overlapping generations are due to the tendency that the former implement a form of local and the latter a form of random replacement.

Even though this exercise shows that natural selection, operating as an external, environmental mechanism, is neither necessary nor sufficient for the creation of clustered populations, we do not want to argue against natural selection as an important mechanism in the biological domain and a substantive driving force in the speciation process. To the contrary, the concept of (natural) selection operating at a global level may provide us with plausible interpretations of the model results, even in disciplines where such interpretations are still lacking. In the words of Dobzhanski (1970, pp. 5–6):

> [...] in biology nothing makes sense except in the light of evolution. It is possible to describe living beings without asking questions about their origins. The descriptions acquire meaning and coherence, however, only when viewed in the perspective of evolutionary development.

References

Araújo, T., & Vilela Mendes, R. (2009). Innovation and self-organization in a multiagent model. *Advances in Complex Systems, 12*(2), 233–253.

Axelrod, R. (1997). The dissemination of culture: A model with local convergence and global polarization. *The Journal of Conflict Resolution, 41*(2), 203–226.

Banisch, S. (2010). Unfreezing social dynamics: Synchronous update and dissimilation. In A. Ernst & S. Kuhn (Eds.), *Proceedings of the 3rd World Congress on Social Simulation (WCSS2010)*, Kassel.

Banisch, S., Lima, R., & Araújo, T. (2012). Agent based models and opinion dynamics as Markov chains. *Social Networks, 34*, 549–561.

Castellano, C., Fortunato, S., & Loreto, V. (2009). Statistical physics of social dynamics. *Reviews of Modern Physics, 81*(2), 591–646.

Crow, F., & Kimura, M. (1970). *An introduction to population genetics theory*. New York: Harper and Row.

Deffuant, G., Neau, D., Amblard, F., & Weisbuch, G. (2001). Mixing beliefs among interacting agents. *Advances in Complex Systems, 3*, 87–98.

Dieckmann, U., & Doebeli, M. (1999). On the origin of species by sympatric speciation. *Nature, 400*(6742), 354–357.

Dobzhanski, T. (1970). *Genetics of the evolutionary process*. New York: Columbia University Press.

Drossel, B. (2001). Biological evolution and statistical physics. *Advances in Physics, 50*(2), 209–295.

Fisher, R. (1930). *The genetical theory of natural selection*. Oxford: The Clarendon Press.

Huberman, B. A., & Glance, N. S. (1993). Evolutionary games and computer simulations. *Proceedings of the National Academy of Sciences, 90*(16), 7716–7718.

Kauffman, S. A. (1993). *The origins of order: Self-organization and selection in evolution*. Oxford: Oxford University Press.

Kehoe, T. J., & Levine, D. K. (1984). Regularity in overlapping generations exchange economies. *Journal of Mathematical Economics, 13*, 69–93.

Kondrashov, A. S., & Shpak, M. (1998). On the origin of species by means of assortative mating. *Proceedings of the Royal Society London B, 265*, 2273–2278.

Korolev, K. S., Avlund, M., Hallatschek, O., & Nelson, D. R. (2010). Genetic demixing and evolution in linear stepping stone models. *Reviews of Modern Physics, 82*, 1691–1718.

Moran, P. A. P. (1958). Random processes in genetics. *Proceedings of the Cambridge Philosophical Society, 54*, 60–71.

Seneta, E. (2006). *Non-negative matrices and Markov chains*. Springer series in statistics (2nd ed.). New York: Springer.

Smith, J. M. (1966). Sympatric speciation. *The American Naturalist, 100*(916), 637–650.

Wright, S. (1932). The roles of mutation, inbreeding, crossbreeding, and selection in evolution. In *Proceedings of the Sixth International Congress on Genetics*.

Chapter 9
Aggregation and Emergence: A Synthesis

This chapter is an attempt to synthesize some of the thoughts that have been developed throughout this work and relate them to computational accounts of emergence. While more specific problems concerning the methods developed in the previous chapters—their limitations and possible generalizations—have been discussed at the end of each chapter, this chapter aims at a synthetic view on how this work may contribute to an important aspect of complexity science. We discuss definitions of weak emergence put forth by Bedau (1997), Huneman and Humphreys (2008) and others and show that a definition of emergence in terms of lumpability provides a link between the concept of dynamical incompressibility and Wimsatt's notion of non-aggregativity (Wimsatt 1986).

9.1 The Computational View on Emergence

To my point of view, one of the most important contributions is the perspective that a Markov chain theory of aggregation for agent-based models (ABMs) may provide on emergence and emergent phenomena. Namely, as sketched at the end of the second chapter, ABMs along with cellular automata (CA), genetic algorithms and other related computational tools play an increasingly important role in the philosophical discussions around emergence. Interestingly enough, some philosophers advocate a position which makes use of computational models as a playground to address fundamental questions of emergence (Bedau 1997; Huneman and Humphreys 2008; Humphreys 2008, among others). Questions about the relation of these artificial model environments to real phenomena are not ignored, but considered as an independent issue which is actually part of another

© Springer International Publishing Switzerland 2016
S. Banisch, *Markov Chain Aggregation for Agent-Based Models*,
Understanding Complex Systems, DOI 10.1007/978-3-319-24877-6_9

debate.[1] In this way, the philosophical controversy that usually comes with the term "emergence" (see O'Connor and Wong 2012) is circumvented to some extend. This work has been following a similar tradition, as the models dealt with are also very simple and the focus has been on the method rather than on empirical adequacy.

Let us quote from the introduction to a special issue on "Dynamical Emergence and Computation" in Minds & Machines (2008) Volume 18 (Huneman and Humphreys 2008, p. 426) in order to point out in what regard our approach may contribute to these discussions:

> Other problems appear concerning the *criteria* for those types of emergence. Up to this point, the link between criteria for emergence and simulation can be put in this way: Is the unavoidability of simulation, or the incompressibility of computing the final state, a sufficient criterion for diachronic emergence? And is this criterion epistemological or ontological, i.e. does this criterion lead to a description of emergence which depends on our cognitive abilities, or does it provide an *essential* characterization of the phenomenon? Yet how can we prove that the unpredictability, except by simulations, picks out an objective property of a system, and is not a peculiar limitation of our cognitive abilities? What kind of further criteria do we need if we want a more fine-grained classification of diachronic emergences? Can we derive such a classification from a typology of simulations?

By means of a formulation of computational models as Markov chains we may shed new light on some of these questions. First of all, a Markov chain formulation at the micro level challenges a definition that makes strong reference to simulation (as by "unavoidability of simulation", Huneman and Humphreys 2008, p. 426) or likewise to analytical unpredictability ("emergent phenomena [...] as unpredictable in an analytical way from the equations of the system", Huneman and Humphreys 2008, p. 425) of dynamical emergent phenomena. All criteria with reference to our capabilities of dealing analytically with the problems are prone to mere epistemological accounts of emergence and will not lead to an ontology of dynamical emergences. On the other hand, the possibility or impossibility to lump sets of states into a new chain at the macro level—that is, to compress the process—indicates that the dynamical incompressibility of a computational process might indeed form an appropriate criterion for dynamical emergence. Though both "the unavoidability of simulations" and "the incompressibility of computing the final state" capture the essential aspect of computational emergence, dynamical incompressibility seems to be the more sophisticated argument.

This is not least so because it should be in the relation between the micro and the macro where appropriate criteria for emergence may be found. In fact, the analysis of the micro-macro link in simple computational models as the voter model (VM), leads naturally to some of the conditions for non-aggregativity (and therefore emergence) proposed in Wimsatt (1986, 2006a). Wimsatt's aggregativity conditions, most importantly inter-substitution of parts, are derived in a probabilistic framework to be the mathematical conditions that have to be met so that the micro process is

---

[1] To me, this position has been most clearly articulated by Humphreys (2012) in a talk given at the 2012 DPG Tagung. See Symons (2008) for a critical account on the possible contributions of this approach to the "metaphysical" problem of emergence.

lumpable. This work suggests to base a proper differentiation between emergent phenomena and non-emergent features on the question whether they meet certain aggregativity conditions, notably, in a dynamical setting.

9.2 Challenging Definitions of Weak Emergence

9.2.1 Unavoidability of Simulations

Bedau (2003) defines weak emergence in the following way:

> Assume that P is a nominally emergent property possessed by some locally reducible system S. Then P is weakly emergent if and only if P is derivable from all of S's micro facts but only by simulation. (p.15)

We have shown in Chap. 3 that many computational models can be formalized as random walks on more or less regular graphs. The configuration space Σ of the model is the set of all possible configurations of the simulation and for a considerable class of models we are able to derive explicitly the transition probabilities between all these configurations, among them models that are widely considered to give rise to weakly emergent patterns. As a result we have at hand an analytical description of the computational model that allows us, in principle, to understand all the dynamical processes, final states or stationary distributions without a need of performing simulations. For instance, if the emergent property P is a stable pattern which the model converges to, this corresponds to an absorbing state in the Markov chain formulation. We are able to compute the probability with which the random walk will end up in that (and in any alternative absorbing) state, and, even without any computation, we know that convergence happens in finite time. In principle, therefore, *simulations are avoidable*.

However, analytical predictability of the model results are challenged by the exponential increase of the dimension of the Markov chain description as the number of elements increases. In practice, therefore, the problem remains unpredictable at this level of description and there is no other choice than performing simulations. The resulting unavoidability of simulations is then essentially due to the cognitive difficulties to derive and calculate the explicit Markov chains for some more complicated and bigger models and the technical impossibility to handle matrices of that size on current computer systems. Seen in this way, a criterion based on the analytical unpredictability of a model of emergent phenomena is an epistemological criterion and leads to an epistemological account of emergence. One could also object that even if an analytical description in form of a micro chain $(\Sigma, \hat{P})$ is found, one still has to "simulate" the chain by applying the transition matrix to a certain initial distribution of interest. One still has to rely on all "micro facts". On the other hand, however, certain properties of the chain (as, for instance, convergence to an absorbing state in finite time) can be assessed without any computation or reduction to a simpler description (see also Izquierdo et al. 2009 who mention some of these properties for rather complex models).

To my point of view, criteria for emergent behavior should not be defined with reference to our capabilities of dealing with the respective problem using analytical methods. It seems that this view originates from an implicit tendency to assume that analytical descriptions are differential equation describing the problem at an aggregate macro level, disregarding the possibility of an analytical formulation at the micro level. For our mathematical tools are under constant development and because we cannot foresee whether new methods for the analysis of complex systems allow analytical predictions in the future, any such criterion makes—by the very construction—emergence a purely epistemological question and rules out any hypothesis about emergent phenomena in an ontological sense.

9.2.2 Computational Incompressibility

Very often, the necessity of performing simulations has been related to the impossibility to reduce the problem to a simpler one by deriving "directly" a macroscopic description of the problem. In his 2003 paper Bedau (2003), from which the above definition has been cited, Bedau himself equates weak emergence and computational irreducibility ("Computational irreducibility—that is, weak emergence", p.18) with reference to the well-known work on CA by Wolfram (1994). Here we will use the term *computational incompressibility* which has been used by Huneman and Humphreys (2008), Humphreys (2008) especially in the context of diachronic (i.e., dynamical) emergence.

While preserving the essence of the argument, the concept of dynamical incompressibility provides a definition of weak emergence which is not by construction an epistemological one, because the question whether a process or a model can be compressed is truly a property of the process. It is certainly closely related to analytical predictability (and thus to the unavoidability of simulations), namely, when a macro description in form of, for instance, a differential equation is considered as the reference analytical formulation to which the micro process can be reduced. According to this view, a property is weakly emergent if the process leading to the generation of it is computationally incompressible (Humphreys 2008). Markov chain aggregation provides precise arguments for whether such a complexity reduction is feasible. The transition from micro to macro corresponds, in essence, to a compression of subsets of micro configurations into macro states and in this way Markov chain aggregation—that is, lumpability—operationalizes the concept of computational compressibility. The question of dynamical incompressibility (and therefore, the question of emergence) is then understood as whether this transition from micro to macro is with or without loss of information.

More precisely, we have seen in the third chapter (and other have before, e.g., Shalizi and Moore 2003) that any system property defines a partition of the state space of the micro chain. Then, there are necessary and sufficient conditions

(Kemeny and Snell 1976, Theorem 6.3.2; see also Sect. 3.3) for the process projected onto this partition to be again a Markov chain, that is, a macro description of the process which contains all information about the system (in the true sense of information, see Chap. 7, Sect. 7.4). If we want to keep up a definition of emergence on the basis of dynamical incompressibility, we could put forth the following definition: *a system property P is emergent if the system is lumpable with respect to the macro description defined by P.*

At its core, this Markov chain approach to emergence deals with a type of emergent behavior that might be called "process emergence". It relates processes at a micro level, typically those arising from simple local rules, to processes at a macro level typically obtained by averaging or aggregating certain properties. The focus is not on the emergence of a stable pattern or a higher level structure that "results" from a simulation, it is rather on establishing a relation between processes on the different levels. In fact, "resultant" stable or recurring macro properties are often reflected, in the process point of view, as absorbing or meta-stable states with a high stationary probability (or classes of those).

Notice that this view on computational incompressibility is closely related to the information-theoretic approaches to measure complexity and emergence as developed, for instance, in Grassberger (1986), Lindgren and Nordahl (1988), Crutchfield and Young (1989), Crutchfield and Shalizi (1999), Shalizi and Crutchfield (2001), Shalizi et al. (2004), Ball et al. (2010), and Gmeiner (2013) and that a construction based on Markovianity is in fact a special case of these more general approaches.

I briefly comment on some possible objections to such a definition. The first point is in fact a general difficulty of defining a property as weakly emergent if the processes leading to it are computationally incompressible. Consider, as an example, the emergence of "spatial alignment" in voter-like and segregation models. See Fig. 9.1 for an instance of "spatial alignment" that emerged in a simple version of Schelling's segregation model (Schelling 1971). In the two-community model (for the VM as well as for the CVM) we observe the emergence of "spatial alignment" or segregation in the form of intra-community consensus and inter-community polarization (Chaps. 5 and 6). However, the process with $2^{(M+L)}$ states at the micro level (M and L being the community sizes) is compressible to a description of size $(M + 1) \times (L + 1)$ which is an *essential reduction*. On the other hand, we have seen that on the ring topology (and similarly for the grid, see Fig. 9.1) the number of states that are needed to obtain a Markovian macro description is larger than $2^N/N$ states which cannot be considered "essentially simpler than the microscopic computational process by which the system's behavior is generated" (Bedau 2003, p. 18). This raises the question whether "spatial alignment" is emergent in one case but not in the other. In the same way one could ask whether the emergence of complete consensus in the VM (which often comes as a surprise to people unacquainted with this kind of models) is emergent if the model is run on a complex network but not emergent if it is run on the complete graph.

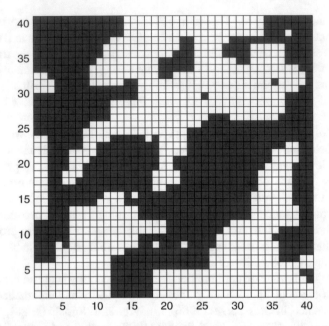

Fig. 9.1 An instance of emergent "spatial alignment" in a simple segregation model

Secondly, a definition based on Markov chain aggregation requires to rigorously define the system property corresponding to a certain emergent feature. On the one hand, this points at a deficiency of existing definitions which are usually not explicit on the property and the processes related to it. On the other hand, it may not always by easy to rigorously define the system property corresponding to a pattern of interest even if the feature catches the observer's eye immediately when looking at the outcome of a simulation. In the above example, one could argue that it is not so clear to which micro configurations an observer would ascribe the property of "spatial alignment". However, there are clearly ways to measure the amount of alignment, for instance, by taking into account the number of aligned neighbors. The respective measure defines again a partition on the space of micro configuration. In fact, pair approximation and approximate mean field theories (Gleeson 2013, and references therein) derive systems of differential equations in a similar way. This point also brings into the discussion the role of the observer and the idea that emergence comes in degrees (Humphreys 2008). In this regard, the non-Markovianity measure used in Sect. 7.4 could be a way to "measure emergence".

Finally, a definition based on exact lumpability may be too unrestrictive on the one and not widely applicable on the other hand. Too unrestrictive because in Markov chains in general lumpability is the exception rather than the rule (Chazottes and Ugalde 2003; Gurvits and Ledoux 2005). Not applicable to important cases, because the full specification of a micro chain and the projection operator may become very complicated for more sophisticated computational models. Even for classical CA with synchronous update, which are widely used in the context of computational emergence, such a specification is by no means straightforward.[2]

9.3   From Computational Incompressibility to Non-aggregativity

Emergence is a concept that relates parts and wholes. Dynamical (or diachronic) emergence is a concept that relates processes on the level of parts to processes on the level of the whole. In our setting, processes at the micro level are processes of change of certain attributes of a set of agents arising from simple local interaction rules. The macro-level process is obtained by the observation or measurement of certain indicators that inform us about the global state of the entire agent population. Typically (though probably not in every case) such a measurement corresponds to an averaging or aggregation of certain agent properties or of properties of the links between agents (i.e., link-between-parts properties). In the VM, for instance, considering the frequency of the different attributes in the population is a one of the simplest forms of aggregation. In the sugarscape model (Epstein and Axtell 1996), to make another example, the number of agents in the environment or the average amount of sugar they carry are similarly aggregative measures. Examples of system properties obtained by aggregation over links-between-parts properties include the number of unaligned neighbors (active bonds) in the VM or, in a more complex setting, the time evolution of density or other structural indicators of a network that co-evolves in the simulation.

Hence, the link between the level of parts and the whole is realized in form of system properties and the typical way to define them is via some form of aggregation. Emergence occurs when the dynamical evolution of the system property ceases to be Markovian. That is, when a Markovian process at the micro-level gives rise to a non-Markovian process at the macro-level. In other words, *when aggregation fails to capture all the dynamical details of the micro-level process.*

In this reading, emergence and "aggregativity" define a fundamental dichotomy between emergent and non-emergent phenomena; a view that, to the authors knowledge, has first been expressed by Wimsatt (1986, 2006a,b):

> *Aggregativity*, a claim that "the whole is nothing more than the sum of its parts", is the proper opposite to *emergence* (Wimsatt 2006b, p. 4)

Calling for an account of emergence consistent with reductionism, Wimsatt starts out from the question of aggregativity: "When intuitively is a system "more than the sum of its parts"?" (Wimsatt 2006a, p. 673). He states four criteria that a system

---

[2]Notice that CA with asynchronous stochastic update, to the contrary, belong to the class of single-step dynamics the specification of which has been the subject of this book.

property in relation with the parts of that system must fulfill to be a fully aggregative property (Wimsatt 2006a, p. 676):

1. invariance of the system property under operations rearranging the parts (inter-substitution),
2. qualitative similarity of the system property under addition or subtraction of parts (size scaling),
3. invariance of the system property under operations involving decomposition and re-aggregation of parts (decomposition and re-aggregation),
4. there are no cooperative or inhibitory interactions among the parts which affect the system property (linearity).

Noteworthy (see Chap. 2), such an account of emergence requires to be explicit on the definition of the system property at question, for one and the same system might be aggregative for one but emergent for another property. This is very similar to lumpability which only makes sense (it is actually only defined) in relation to a certain partition of the state space.

   With the analysis of the VM in Chap. 4 we recover the first of Wimsatt's criteria.[3] Actually we came to the conclusion that aggregativity depends exclusively on the invariance of the system property with respect to the inter-substitution of parts. If the system property (the projection from micro to macro) is defined in accordance with the symmetries of the agent relations, it will be an aggregative measure and correctly describe the evolution of the system (Chap. 5). It might be that models with more complicated interaction mechanisms will require a closer inspection of the other three criteria. On the other hand, Wimsatt (2006a) notes that "[t]hese conditions are not independent of one another." (p.675). Moreover, for all models in which the local transitions depend only on the neighborhood configurations, invariance of agent relations with respect to agent permutations is sufficient for lumpability and ensures compressibility of the process.

   Finally, in order to stress that we are dealing with *dynamical* emergence, let me mention a subtlety when applying Wimsatt's arguments to the VM. One could actually argue that all of the aggregativity criteria are met by a macro formulation in terms of attribute frequencies $h(\mathbf{x})$ (Hamming weight) independent of the agent network. Namely, at each instance, that is, for each single micro configuration, the system property $h(\mathbf{x})$ satisfies all of Wimsatt's aggregativity conditions. It is invariant with respect to alternative arrangements of the agents. Qualitative similarity (even an obvious quantitative relation) after addition or subtraction of parts is also satisfied. Thirdly, given any configuration $\mathbf{x}$, it is possible to decompose the system in an arbitrary way, compute $h(\mathbf{x})$ for subsets of agents, and re-aggregate (sum up) the measures for the different subsets. Finally, $h(\mathbf{x})$ is also invariant with

---

[3]One of Wimsatt's main concerns is to show that in natural phenomena full aggregativity (present if all four conditions are satisfied) is the exception rather than the rule and that in many models—including voter-like models of population genetics—the use of aggregative procedures is unjustified (Wimsatt 2006a).

respect to non-symmetric interaction relation, just because it only takes into account node (agent) properties. Noteworthy, this works for any agent network.

The answer to this simple puzzle resides in the shift from a synchronic to a diachronic—that is, process-based—perspective. Namely, in the context of Markov chain aggregation the system property on which aggregativity conditions must be assessed is not the attribute frequency, but rather the transition probability from one frequency level to the other. From the point of view of lumpability this is obvious, but it is not, in general, from the point of view of emergence. A dynamical argument is also presented in Wimsatt (2006a) to illustrate aggregation failures in classical population genetics, but the distinction between a synchronic and a diachronic emergence is not always clear. It might even be the case (but this certainly deserves a further inspection) that the aggregativity criteria with reference to dynamical operations (most importantly criteria 4) appear redundant if an explicit process perspective is taken.

To sum up, a definition of emergence in terms of lumpability provides a link between the concept of dynamical incompressibility and Wimsatt's notion of non-aggregativity. It shares an intrinsic emphasis on processes with the former and with the latter a clear concept of system property as well as the idea that emergence and aggregativity define a dichotomy between emergent and non-emergent phenomena.

References

Ball, R. C., Diakonova, M., & Mackay, R. S. (2010). Quantifying emergence in terms of persistent mutual information. *Advances in Complex Systems, 13*(03), 327–338.

Bedau, M. A. (1997). Weak emergence. *Philosophical Perspectives, 11*, 375–399.

Bedau, M. A. (2003). Downward causation and the autonomy of weak emergence. *Principia Revista Internacional de Epistemologica, 6*(1), 5–50.

Chazottes, J.-R., & Ugalde, E. (2003). Projection of Markov measures may be Gibbsian. *Journal of Statistical Physics, 111*(5/6), 1245–1272.

Crutchfield, J. P., & Shalizi, C. R. (1999). Thermodynamic depth of causal states: Objective complexity via minimal representations. *Physical Review E, 59*(1), 275.

Crutchfield, J. P., & Young, K. (1989). Inferring statistical complexity. *Physical Review Letters, 63*(2), 105–108.

Epstein, J. M., & Axtell, R. (1996). *Growing artificial societies: Social science from the bottom up*. Washington, DC: The Brookings Institution.

Gleeson, J. P. (2013). Binary-state dynamics on complex networks: Pair approximation and beyond. *Physical Review X, 3*, 021004.

Gmeiner, P. (2013). Some properties of persistent mutual information. In *Proceedings of the European Conference on Complex Systems 2012* (pp. 867–876). Berlin: Springer.

Grassberger, P. (1986). Toward a quantitative theory of self-generated complexity. *International Journal of Theoretical Physics, 25*(9), 907–938.

Gurvits, L., & Ledoux, J. (2005). Markov property for a function of a Markov chain: A linear algebra approach. *Linear Algebra and Its Applications, 404*(0), 85–117.

Humphreys, P. (2008). Synchronic and diachronic emergence. *Minds and Machines, 18*(4), 431–442.

Humphreys, P. (2012). Ising models: Interpretational and computational issues. In *Jahrestagung der DPG 2012 in Berlin, Arbeitsgruppe Philosophie der Physik*.

Huneman, P., & Humphreys, P. (2008). Dynamical emergence and computation: An introduction. *Minds and Machines, 18*(4), 425–430.

Izquierdo, L. R., Izquierdo, S. S., Galán, J. M., & Santos, J. I. (2009). Techniques to understand computer simulations: Markov chain analysis. *Journal of Artificial Societies and Social Simulation, 12*(1), 6.

Kemeny, J. G., & Snell, J. L. (1976). *Finite Markov chains*. New York: Springer.

Lindgren, K., & Nordahl, M. G. (1988). Complexity measures and cellular automata. *Complex Systems, 2*(4), 409–440.

O'Connor, T., & Wong, H. Y. (2012). Emergent properties. In E. N. Zalta (Ed.), *The Stanford encyclopedia of philosophy* (Spring 2012 edn.). http://plato.stanford.edu/archives/spr2012/entries/properties-emergent/.

Schelling, T. (1971). Dynamic models of segregation. *Journal of Mathematical Sociology, 1*(2), 143–186.

Shalizi, C. R., & Crutchfield, J. P. (2001). Computational mechanics: Pattern and prediction, structure and simplicity. *Journal of Statistical Physics, 104*(3–4), 817–879.

Shalizi, C. R., & Moore, C. (2003). What is a macrostate? Subjective observations and objective dynamics. In *CoRR*. arXiv:cond-mat/0303625.

Shalizi, C. R., Shalizi, K. L., & Haslinger, R. (2004). Quantifying self-organization with optimal predictors. *Physical Review Letters, 93*, 118701.

Symons, J. (2008). Computational models of emergent properties. *Minds and Machines, 18*(4), 475–491.

Wimsatt, W. C. (1986). *Forms of aggregativity* (pp. 259–291). Dordrecht: Reidel.

Wimsatt, W. C. (2006a). Aggregate, composed, and evolved systems: Reductionistic heuristics as means to more holistic theories. *Biology & Philosophy, 21*(5), 667–702.

Wimsatt, W. C. (2006b). Reductionism and its heuristics: Making methodological reductionism honest. *Synthese, 151*(3), 445–475.

Wolfram, S. (1994). *Cellular automata and complexity: Collected papers* (Vol. 1). Reading, MA: Addison-Wesley.

Chapter 10
Conclusion

This book presents a Markov chain approach for the analysis of agent-based models (ABMs). It provides a general framework of aggregation in agent-based and related computational models by making use of Markov chain aggregation and lumpability theory in order to link between the micro-level dynamical behavior and higher-level processes defined by macroscopic observables. The starting point is a formal representation of a class of ABMs as Markov chains— so called micro chains— obtained by considering the set of all possible agent configurations as the state space of a huge Markov chain. This allows for the application of the theory of Markov chain aggregation—namely, lumpability—in order to reduce the state space of the models and relate microscopic descriptions to a macroscopic formulation of interest. In some cases, the aggregation is without loss of information and the macro chain can be used to compute several stationary and transient characteristics of the models. In general, however, a certain amount of macroscopic complexity is introduced by the transition from the micro level to a favored macro description which is a fingerprint of emergence in agent-based computational models.

1. *ABMs are Markov chains.*

While Markov chains represent a relatively simple form of a stochastic process, ABMs put a strong emphasis on heterogeneity and social interactions. Nevertheless, *most ABMs are from the formal point of view Markov chains.* Intuitively, this might be clear by the fact that ABMs usually come in form of a computer program which takes a certain initial population of agents as an input and iteratively applies an algorithm to evolve the agent population from one time step to the other. In order to formally represent such an iterative process as a Markov chain, a single state of the chain must be conceived of as a possible configuration of the entire system and contain all the dynamical variables and microscopic details—agent attributes, their connectivity structure, state of the environment etc. A rigorous proof of the Markovianity of ABMs is not always straightforward. However, the micro process is a Markov chain whenever the iteration it implements can be understood as a

© Springer International Publishing Switzerland 2016
S. Banisch, *Markov Chain Aggregation for Agent-Based Models*,
Understanding Complex Systems, DOI 10.1007/978-3-319-24877-6_10

(time-independent) stochastic choice out of a set of deterministic options. In this respect, the random mapping representation of a Markov process (Chap. 3, Sect. 3.2) helps to understand the role of the collection of (deterministic) dynamical rules used in the model from one side and of the probability distribution ω governing the sequential choice of the dynamical rule used to update the system at each time step from the other side.

2. *A class of ABMs that we have called single-step models give rise to random walks on regular graphs.*

Moreover, for a class of models which we have referred to as *single-step dynamics*, it is possible to derive explicitly the transition probabilities $\hat{P}$ in terms of the update function **u** and the probability distribution ω. Due to a sequential update mechanism in which an agent along with a set of neighbors is chosen and the state of that agent is updated as a function of the neighborhood configuration, non-zero transition probabilities are possible only between configurations that differ in at most on element (one agent). *This characterizes ABMs as random walks on regular graphs.* Namely, in a system with N agents each of which may be in one out of δ states, the set Σ of all agent configurations is the set of strings of length N of δ possible symbols. Under sequential update of only one agent at a time, transitions are possible only between adjacent strings so that the maximal "grammar" of such a system is the Hamming graph $H(N, \delta)$. However, a completely regular walk on $H(N, \delta)$ with non-zero transition probabilities between all adjacent configurations is realized only if no constraints act in the system. In particular, as will be resumed below, if the interaction probabilities and therewith the distribution ω are constrained, for instance, by an underlying interaction network, the structure of the micro chain becomes more and more irregular. The same is true for other interaction constraints such as assortativity or bounded confidence.

3. *Regularity implies dynamical redundancy and therefore the possibility of state space reduction.*

Nevertheless, the approach to AB simulations as random walks on more or less regular graphs hints at the possibility of reducing the state space of the micro chain by exploiting systematically the dynamical symmetries that an ABM gives rise to. Namely, the existence of non-trivial automorphisms of the micro chain tells us that certain sets of micro configurations can be interchanged without changing the probability structure of the chain. *These sets of micro states can be aggregated or lumped into a single macro state and the resulting macro-level process is still a Markov chain.* In Markov chain theory, such a state space reduction by which no information about the dynamical behavior is lost is known as lumpability.

4. *Macro observations and system properties induce state space partitions and reductions.*

There is another way of looking at state space reductions or aggregation which is particularly relevant in the study of ABMs. Namely, *any observable of the system naturally defines a many-to-one relation by which sets of micro configurations with the same observable value are aggregated into the same macro state.* In other words, tracking the time evolution of a model in terms of a system property or order parameter ϕ that informs us about the global state of the system corresponds to a projection Π of the micro chain onto a partition $\mathbf{X}$ of the space of micro configurations Σ. Vice versa, any projection map Π from Σ to $\mathbf{X}$ defines an observable ϕ with values in the image set $\mathbf{X}$ that are in complete correspondence with a classification based on an observable property of the system. These two ways of describing the construction of macro-dynamics are equivalent and the choice of one or the other point of view is just a matter of taste.

5. *A macro observation defines a Markov process if it is compatible with the symmetries of the micro chain.*
6. *Vice versa, the symmetries of the micro chain induce a partition with respect to which the process is lumpable*

The main question that this thesis has been concerned with is about the conditions on the microscopic system $(\Sigma, \hat{P})$ and the projection construction $(\Pi : \Sigma \to \mathbf{X}$ or respectively ϕ) that have to be met in order to lead to a macro process that is still a Markov chain. The starting point has been Kemeny and Snell (1976, Theorem 6.3.2), in which necessary and sufficient condition for lumpability are provided. On that basis, a sufficient condition has been provided with Theorem 3.2 that relates the question of lumpability to the symmetries in the dynamical structure of the micro chain. Namely, to any partition $\mathbf{X}$ of Σ there is a transformation group $\mathcal{G}$ acting on Σ that generates $\mathbf{X}$ and Theorem 3.2 states that the micro process $(\hat{P}, \Sigma)$ is lumpable to a macro process $(\mathbf{X}, P)$, if $\hat{P}$ is symmetric with respect to $\mathcal{G}$. The automorphisms of the microscopic transition matrix (for $\mathcal{G} \subseteq Aut(\hat{P})$, see Lemma 3.1) can therefore be used to construct a partition $(\mathbf{X})$ with respect to which the process is lumpable. In turn, an observation on the system will define a lumpable macro process if it is compatible with the symmetries of the micro chain.

7. *In the voter model, homogeneous mixing is a prerequisite for lumpability with respect to aggregation over all agent attributes.*

This thesis has made extensive use of the voter model (VM)—one of the simplest ABMs—to illustrate these points. In the binary VM each agent can be in two possible states □ and ■. At each time step, two agents (linked in the interaction network) are chosen at random with probability $\omega(i, j)$ and one of them copies the state of the other. From the microscopic perspective the binary VM is a random walk on the N-dimensional hypercube $(H(N, 2))$ and the Hamming weight $\phi(\mathbf{x}) = h(\mathbf{x})$ (to maintain this terminology) of an agent configuration is the most typical macro level of observation. In effect, all micro configurations with the same Hamming weight are mapped into the same macro state which is a tremendous reduction from 2^N micro states to $N + 1$ macro states. However, we have shown (Chap. 4, Sect. 4.1.1) that the symmetries of the micro chain $\hat{P}$ are compatible with that level of observation only if the probability distribution ω is invariant with respect to all agent permutations. Markovianity at the macro level requires that the probability with which two agents are chosen $\omega(i, j)$ is equal for all agents pairs which renders homogeneous mixing a prerequisite for lumpability. The resulting process is known as Moran process (after Moran 1958).

8. *The use of Markov chain theory enables a complete characterization of the dynamical behavior of the VM with homogeneous mixing.*

Throughout Chap. 4 the VM with homogeneous mixing and the resulting macro chain on $\mathbf{X} = (X_0, \ldots, X_k, \ldots X_N)$ with $h(\mathbf{x}) = k$ has been discussed in detail. Due to the structure of the macro chain it has been possible to derive a closed-form expression for the fundamental matrix $\mathbf{F}$ for arbitrary N. Encoding the recurrence and hitting times of the system, this provides all the information about the mean quantities and variances of the transient dynamics in this model. Noteworthy, Markov chain theory allows for some computations that are not easy with other methods such as mean-field approaches. For instance, it is possible to characterize the convergence behavior of realizations that end up in one absorbing state independently from those that end up in the other one (Sect. 4.2.4). Moreover, the multi-state VM in which agents can adopt δ different states is shown to be reducible to the binary VM by a further lumping. However, *only if the interaction is unconstrained in the sense that all agents and all attributes interact in the same way.* In turn, if interactions are constrained by assortativity or bounded confidence this may lead to a stable pattern of polarization at the level of the entire population (Sect. 4.2).

9. *Microscopic heterogeneity translates into dynamical irregularities in the micro chain and requires a refinement of the aggregation and the corresponding level of observation.*

When inhomogeneities are introduced in the model, the symmetry conditions for lumpability in Theorem 3.2 (as well as Kemeny and Snell 1976, Theorem 6.3.2,) are no longer satisfied for the partition $\mathbf{X}$ induced by aggregation over all agents (i.e., by $h(\mathbf{x})$). However, Chap. 5 shows at the example of the VM that a refinement of the aggregation procedure is possible which is based entirely on the symmetries of the interaction probabilities ω. Proposition 5.1 states that the automorphism of ω may be used to define a group $\mathscr{G}$ of bijections on Σ which generates a lumpable partition $\mathscr{M}_\omega$ of Σ. Noteworthy, while Theorem 3.2 as well as common approaches to lumpability require the analysis of the δ^N-dimensional micro chain, with Proposition 5.1 the result is stated in terms of the symmetries of a interaction network of size N. The most important implication of Proposition 5.1 is that the higher the amount of heterogeneity in the agent system, the lesser the coarse-graining that is possible if a Markovian description is desired to capture all the dynamical details of the micro process. In other words, *the more constrained and heterogeneous the microscopic interaction probabilities and rules, the more irregular the micro process and the lower the chances to obtain a reasonable reduction by Markov chain aggregation.*

10. *Markov chain aggregation leads to solvable chains (only for) for "simple" population structures.*

It is clear then that the exact aggregation by lumpability significantly reduces the number of states only if the interaction network that underlies the model possesses a lot of symmetries. This restricts the applicability of the method as a solution technique for ABMs to stylized situations as the leader-follower topology (Sect. 5.5) or the two-community model (Sects. 5.4 and 6.4). Nevertheless, even in those stylized situations interesting features can be observed. In the leader-follower system, the probability that the leader imposes its opinion on a follower population increases with the influence asymmetry between a leader and the followers but is independent of peer-to-peer processes among followers. A greater influence of the leader also accelerates the convergence process, however, this effect is undermined by a stronger peer-to-peer communication (Sect. 5.5). For the two-community VM, in which a weak influence exists between two strongly connected groups, a general increase in convergence times is observed due to the existence of meta-stable states of intra-community consensus and inter-community polarization. This is also observed in the quasi-stationary distribution of the two-community VM.

Similar results are obtained for the contrarian VM (CVM) in which agents act in a contrarian way with a small probability p (Chap. 6). An increasing contrarian rate leads to a process that is characterized more and more by random state flips independent of whether the agents are completely connected or organized in communities. As p becomes smaller, topological effects become visible. On the complete graph the population is almost uniform for long periods of time, but due to the random perturbations introduced by the contrarian rule there are rare transitions between the two consensus profiles. Noteworthy, there is a characteristic p at which the rate of switching becomes maximal (Sect. 6.3.2). On the two-community graph the system is likely to approach the states of inter-community polarization and remain there for quite some time. Such ordering behavior is also observed for other networks with a strong local structure (Sect. 7.2.2).

11. *Microscopic heterogeneity leads to macroscopic complexity.*

Another interpretation that Proposition 5.1 suggests is that microscopic hetero-geneity introduces complexity at the macroscopic level. This idea has been taken up in Chap. 7 using the CVM as an example. If we decide to stay at the level of full aggregation over all agents ($h(\mathbf{x})$) despite the fact that it is not compatible with the symmetries of the micro chain, the process obtained by this projection is no longer a Markov chain. This means that a certain amount of memory is introduced at the macroscopic level by the very way the system is observed. In the last part of Chap. 7, this divergence from Markovianity has been quantified in terms of the information that the past (before the present) contains about the future. The two-community CVM has served as a scenario in which these entities can be explicitly computed. Again, there is a characteristic contrarian rate p at which deviations from Markovianity are maximal.

The method informs us in this way about the complexity of a system intro-duced by non-trivial interaction relations. Namely, the theory of Markov chain aggregation makes explicit statements about when a micro process is *compressible* to a certain macro-level description. This links non-lumpability to computational incompressibility, one of the key concepts in dynamical emergence (Bedau 2003; Huneman and Humphreys 2008, among others). Moreover, in the context of Markov chain aggregation, computational incompressibility becomes directly related to Wimsatt's notion of non-aggregativity (Wimsatt 1986, 2006), another important account of emergence. The argumentation in Chap. 9 suggests that deviations from Markovianity at the macro level can be understood as a fingerprint of dynamical emergence—and hence complexity—as the macroscopic process displays features that are not present in the micro level process.

The models used throughout this book are very simple and I would not claim that they are reasonable descriptions of real social phenomena. Their main purpose is to shed light on some fundamental mechanism of self-organizing systems. In this regard, I would like to emphasize the role that constraints on the agent behavior play regarding the aggregativity or reducibility of the models to a macro-level description. Even in those simple models, complex and heterogeneous interactions structures rule out completely the possibility of deriving a loss-less Markovian macro description which is sensitive to all dynamical details. Likewise do constraints on the interaction rules (as assortativity) necessitate the inclusion of more detail into a valid macro-level description in order to account for population effects (as polarization) that emerge from them. With their obvious limitations, the models used here do not allow for a direct generalization to more realistic cases, their treatment is only the first step in the stochastic analysis of the micro-macro link in social simulation. On the other hand, an increase in model complexity in more sophisticated ABMs comes often by introducing various levels or dimensions of agent heterogeneity, different types of agents with different rules, and a possibly variable environment. The macro patterns we observe in them are always the result of an adjustment of the constraints on and heterogeneities in the microscopic system and the fundamental mechanisms those that are at play also in the simpler models.

Nevertheless, the application of the ideas presented in this thesis to other ABMs is certainly an interesting issue for the future. Even if this has to be carefully considered model by model, a micro formulation in form of a micro chain will usually be possible (see Sects. 3.2.3 and 4.3). For more sophisticated ABMs, however, deriving an exact aggregate description of a size which allows for direct computations by the use of strong lumpability is not likely. On the other hand, there are nowadays powerful computational techniques to deal with large Markov chains, and interestingly, these methods are often based on approximate aggregations of the chains (Buchholz 2006; Stewart 2009; Touzene 2013, among many others). In this context, the presented concepts might help in the analysis of the adequacy of such approximate techniques, and they may also shed light on the relation between approximate aggregations and the macroscopic measures they can be associated with.

The method is most directly applicable to models at use in socio-cultural dynamics, evolutionary graph theory as well as to stochastic cellular automata (CA). Regarding the first field, one interesting extension concerns simple forms of memory in the agent decisions such as agents that remember the states that they have already visited (Bornholdt et al. 2011) or by assuming that the probability of an agent to change its opinion decreases with the time it sticks to the current one (Stark et al. 2008). The macroscopic effects of these simple extensions are very interesting and encouraging for further analysis. Likewise, a more sophisticated modeling consists of coupling the individual agent dynamics with the macro dynamics and allow certain macro-structural properties to feed back onto the level of individual decisions. Such ingredients have been introduced into models of herd behavior in finance (Krause and Bornholdt 2013) and they are also relevant in voting behavior (Caruso and Castorina 2005).

Secondly, the main question in evolutionary graph theory is how the population structure ω affects the outcome of an evolutionary process. It is now well-known that certain population structures may enhance or suppress selection in the sense that the probability of a randomly placed mutant to invade the entire population differs from the respective Moran probability obtained for homogeneous mixing (Liberman et al. 2005). We have seen such a divergence from the Moran probabilities in Sect. 5.5 where the VM on a leader-follower topology has been discussed. The search for paradigmatic network structures which affect the fixation probabilities is still a topic of current research (e.g., Shakarian et al. 2012; Voorhees and Murray 2013) and the methods developed here, Proposition 5.1 in particular, can be directly applied to study not only exit probabilities, but also the pace of mutant fixation.

Thirdly, regarding the idea of relating lumpability to dynamical emergence (Chap. 9), the application to CA is a particularly compelling idea. However, a micro description of the original synchronous CA as Markov chains is challenging as, in principle, all agents can change at a time (notice that original CA are deterministic systems, but that their transitions can nevertheless be encoded in a "transition matrix"). On the other hand, their asynchronous probabilistic counterparts belong precisely to the class of single-step dynamics which we have been concerned with in this book. (See, for instance, Schönfisch and de Roos, 1999; Nehaniv 2004 for the relation between asynchronous and synchronous automata.) In particular, when the probability of choosing a triple (i, j, k) of cells is equal for all triples (complete graph), the micro chain is lumpable with respect to a (macro) description in term of the number of white and respectively black cells (as sensible as such a description might be). Some preliminary computations with the respective macro chains indicate, that the different rules alone, even in a non-localized form, lead to behaviors by which the more complex rules are distinguishable from the simpler ones. This might be useful for classification. While this work has been more concerned with the effects of heterogeneity in ω, the systematic study of elementary CA in a homogeneous setting could be a way to understand the contribution of different update rules **u** to the dynamic behavior of complex computational models.

Another more general topic that should be addressed in the future is to obtain a more detailed but also more synthetic understanding of the macroscopic effects that may emerge in micro simulation models. One starting point could be a quantification of the range of memory at the macro level in order to gain insight about the microscopic conditions for long-term memory effects that are known to exist in many real world systems from Finance (Cont 2001) to Biology (Stanley et al. 1994). This work shows that a crucial role in the emergence of temporal correlations is played by the patterns of microscopic heterogeneity implemented in the agent model, the nature of the correlations due to the underlying structure and the constraints in the interaction rules. Along the lines of Chap. 7, insight into the fundamental principles for the emergence of temporal correlations in microscopic simulation models could be obtained by applying information-theoretic tools and Markov chain theory to ABMs.

More generally, under certain circumstances the macro process may undergo dynamical changes in its own structural rules. This fact is referred to as explanatory emergence, a controversial issue in social theory (Giesen 1987). It can be understood either as a consequence of some external (to the model) inputs or on the basis of deep accelerations of the micro dynamics that in turn bring about the processes of change at the macro level. In both cases this question opens up to new theoretical as well as very interesting practical developments.

All in all, the theory of Markov chain aggregation applied to ABMs provides a useful instrument for the analysis of the link from a microscopic AB description of a dynamical system to macroscopic observables and may therefore contribute to our understanding of aggregation and emergence in complex adaptive systems.

References

Bedau, M. A. (2003).   Downward causation and the autonomy of weak emergence. *Principia Revista Internacional de Epistemologica, 6*(1), 5–50.

Bornholdt, S., Jensen, M. H., & Sneppen, K. (2011).   Emergence and decline of scientific paradigms. *Physical Review Letters*, 106(5), 058701.

Buchholz, P. (2006). Structured analysis techniques for large Markov chains. In *Proceeding from the 2006 Workshop on Tools for Solving Structured Markov Chains*, SMCtools '06. ACM: New York, NY.

Caruso, F., & Castorina, P. (2005).   Opinion dynamics and decision of vote in bipolar political systems. *International Journal of Modern Physics C, 16*(09), 1473–1487.

Cont, R. (2001).   Empirical properties of asset returns: Stylized facts and statistical issues. *Quantitative Finance, 1*(2), 223–236.

Giesen, B. (1987). Beyond reductionism: Four models relating micro and macro levels. In J. C. Alexander, B. Giesen, R. Münch, & N. J. Smelser (Eds.), *The micro-macro link* (Chapter 15). Berkeley: University of California Press.

Huneman, P., & Humphreys, P. (2008). Dynamical emergence and computation: An introduction. *Minds and Machines, 18*(4),425–430.

Kemeny, J. G., & Snell, J. L. (1976). Finite Markov chains. In *Informal introduction to stochastic processes with maple*. New York: Springer.

Krause, S. M., & Bornholdt, S. (2013). Spin models as microfoundation of macroscopic market models. *Physica A: Statistical Mechanics and its Applications, 392*(18), 4048–4054.

Liberman, E., Hauert, C., & Nowak, M. (2005).   Evolutionary dynamics on graphs. *Nature, 433*(7023), 312–316.

Moran, P. A. P. (1958). Random processes in genetics. *Proceedings of the Cambridge Philosophical Society, 54*, 60–71.

Nehaniv, C. L. (2004). Asynchronous automata networks can emulate any synchronous automata network. *International Journal of Algebra and Computation, 14*(05n06), 719–739.

Schönfisch, B., & de Roos, A. (1999). Synchronous and asynchronous updating in cellular automata. *Biosystems, 51*(3), 123–143.

Shakarian, P., Roos, P., & Johnson A. (2012). A review of evolutionary graphs theory with applications to game theory. *Biosystems, 107*, 66–80.

Stanley, H., Buldyrev, S., Goldberger, A., Goldberger, Z., Havlin, S., Mantegna, R., et al. (1994). Statistical mechanics in biology: How ubiquitous are long-range correlations? *Physica A: Statistical Mechanics and its Applications, 205*(1), 214–253.

Stark, H.-U., Tessone, C. J., & Schweitzer, F. (2008). Decelerating microdynamics can accelerate macrodynamics in the voter model. *Physical Review Letters, 101*(1), 018701.

Stewart, W. J. (2009). *Probability, Markov chains, queues, and simulation: The mathematical basis of performance modeling*. Princeton: Princeton University Press.

Touzene, A. (2013). A new parallel algorithm for solving large-scale Markov chains. *The Journal of Supercomputing, 67*(1), 239–253.

Voorhees, B., & Murray, A. (2013). Fixation probabilities for simple digraphs. *Proceedings of the Royal Society A: Mathematical, Physical and Engineering Science, 469*, 2154.

Wimsatt, W. C. (1986). *Forms of aggregativity* (pp. 259–291). Dordrecht: Reidel.

Wimsatt, W. C. (2006). Aggregate, composed, and evolved systems: Reductionistic heuristics as means to more holistic theories. *Biology & Philosophy, 21*(5), 667–702.

Printed in the United States
By Bookmasters